The Fiction of Bioethics

Reflective Bioethics

Series Editors:
Hilde Lindemann Nelson and James Lindemann Nelson

The Patient in the Family
Hilde Lindemann Nelson and James Lindemann Nelson

Do We Still Need Doctors?
John D. Lantos, M.D.

Stories and Their Limits: Narrative Approaches to Bioethics
Hilde Lindemann Nelson

Physician Assisted Suicide: Expanding the Debate
Rosamond Rhodes, Margaret P. Battin, Anita Silvers, eds.

A Philosophical Disease
Carl Elliott

Duty and Healing:
Foundations of a Jewish Bioethic
Benjamin Freedman

Meaning and Medicine:
A Reader in the Philosophy of Health Care
James Lindemann Nelson and Hilde Lindemann Nelson, eds.

The Fiction of Bioethics
Cases as Literary Texts
Tod Chambers

Routledge
New York and London

Published in the United States of America in 1999 by
Routledge
29 West 35th Street
New York, NY 10001

Published in Great Britain by
Routledge
11 New Fetter Lane
London EC4P 4EE

Text Design by Tara Klurman

10 9 8 7 6 5 4 3 2 1

LIBRARY OF CONGRESS CATALOGING-IN-PUBLICATION DATA

Chambers, Tod
    The fiction of bioethics: cases as literary texts / Tod Chambers.
    p.   cm.
    ISBN 0-415-91988-6 (hb). — ISBN 0-415-91989-4 (pb)
    1. Bioethics—Case studies.   I. Title.
R725.5.C46 to 1999
174'.2—DC21   98-43178
                                                           CIP

For Megan

"Why couldn't the world that concerns us—be a fiction?"

—*Friedrich Nietzsche*
BEYOND GOOD AND EVIL

Contents

Contents

Acknowledgments

I would like to thank the scholars who have read and commented on the manuscript for this book. First and foremost, I wish to express special thanks to Kathryn Montgomery, who has provided invaluable assistance both in the content and the form of this book; she has patiently read the entire manuscript far too many times to count. I would also like to thank the following individuals who have read either the entire manuscript or significant portions of it: Carl Elliott, Rita Charon, Suzanne Poirier, Hilde Lindemann Nelson, Marilyn Chandler McEntyre, and James Lindemann Nelson.

Preface

This book is primarily concerned with the issue of representation and applied ethics. How real-life situations are portrayed has been an important and rich topic in the social sciences and the humanities; during my graduate training the "cracks in the mirror" for such disciplines as anthropology, sociology, history, economics, and philosophy had already become common knowledge. When I began work in medical ethics, I was surprised to find that the field had largely ignored the way representation affected the examination of moral problems. Each day I found myself in the midst of people testing their ideas through ethical quandaries, but these highly insightful individuals seemed unaware that they were applying their theories to stories and not to unmediated reality. The self-reflexive turn in many of our academic disciplines was an important move toward intellectual honesty, an attempt to come to terms with their rhetoric. I saw that bioethics needed to undergo this turn as well and this book is, I hope, a helpful step in that direction.

A word should be said concerning the approach I use in analyzing this "data" of bioethics. I have no particular theoretical ax to grind in this book, and I view various narrative theories as tools to understand the data. I have been relatively catholic when it comes to theory, and I take liberally from many narrative perspectives. Some purists may be surprised to find that I use theories from structuralism, semiotics, post-structuralism, New Criticism, discourse analysis, media studies, reader-response criticism, and feminism. I admit to a particular tendency in this book toward structuralist methods, but this is not as much a wish to promote this method over others (an argument that would seem odd in the face of recent theory) but for a pragmatic reason: It tends to work well in the analysis of bioethics cases. The reason that these various theories work (and others do not work as well) has more to do with the nature of the bioethics case than the strength of the theories. That I find simple binary splits within bioethics cases represents the simplicity of the traditional case narrative rather than anything necessarily about the universality of a binary world. The sociologist Howard Becker uses the metaphor of woodworking to explain what he considers a good use of "what others have done" (142). After deciding to make a table, you find that you need not make all the parts because some of the parts are of a standard size and you can simply buy them and fit them into place. In a similar way, when analyzing something one should "know the literature" because it saves you the trouble of having to create an entire new theory to explain what is going on. So, for example, I do not need to create categories for different narrative points of view because Gérard Genette has already provided a useful model that can be applied to ethics cases.

Along with ethics cases, I have tried to illustrate the various aspects of narrative by drawing on a variety of examples. Like my selection in theory, my selection of narrative examples tends to be quite broad. The reason for using examples from not only canonical literature (like Joyce and Fitzgerald) but also from detective novels, televisions shows, action comics, and popular movies is to display the "ordinariness" of narrative. Not only does the novelist Vladimir Nabokov distort time in his narratives but so does the action film director John Woo. I do not consider ethics cases to be examples of great literature but rather that in order to understand them as the data of bioethics we must be aware of their form of representation. They are stories.

Chapter One: Stories as Data

"There is no longer any such thing as fiction or nonfiction; there's only narrative."

—*E. L. Doctorow*

THE LEGACY OF THE HANDMAIDEN

One of the recent developments in the field of bioethics has been an engagement with narrative. While there is considerable diversity in what is usually referred to as "narrative ethics" (see Jones "Literature"; Hilde Lindemann Nelson), much of the discussion has conceived of narrative as a handmaiden to philosophy. Both bioethicists and literary scholars have come to appreciate that the subject and analytical tools of the discipline of literature-and-medicine assist in resolving moral problems within health care.

Early on medical ethicists and literature-and-medicine scholars discovered that literature could provide a source for richly rendered ethics cases. Traditional ethics cases have often been criticized for being too thinly rendered to provide an adequate resource for analysis and thus drawing upon literature

1

has been advocated by such scholars as Howard Brody, Joanne Trautmann Banks, Kathryn Montgomery Hunter, and Anne Hudson Jones. As James Terry and Peter Williams summarize this perspective: "Short stories and poems that are evocative, complex, and imaginatively challenging have been used to supplement or supplant the traditional case study as instruments for raising ethical issues.... Philosophic understanding of a given moral problem can be enriched by a literary account that places issues in a context of the lives and activities of particular characters" (1). It has become common for medical students to use the short stories of William Carlos Williams and Richard Selzer to explore issues of informed consent, refusal of treatment, and active euthanasia. And Jones, in her study of Perri Klass's *Other Women's Children*, has demonstrated how entire novels can provide a depth of understanding for moral issues in medicine that is missing from most discussions. Related to this use of literature as rich case study has been the work of physicians such as Robert Coles, who in *The Call of Stories* sees that literature can assist in the development of moral imagination (cf. Radwany and Adelson). Edmund Pellegrino in a similar manner observes that reflecting on literature permits a physician to further develop moral character ("Look").

Beyond the actual subject matter of study, literature-and-medicine scholars have also seen that the critical methods of narrative studies can assist the analysis of moral problems. In "Literature and Medicine: Contributions to Clinical Practice," the authors, leaders in the field of literature-and-medicine, acknowledge that literary works provide an "unequaled" source for ethics cases, but they contend that, "Perhaps more fundamental to ethics than individual literary texts are literature's methods" (Charon et al. 602). Literature-and-medicine scholars have drawn upon the techniques developed in interpreting literary texts in order to facilitate understanding moral problems in medicine. Rita Charon argues that there are stages in the interpretation of ethics cases that are parallel to those used in understanding any narrative: recognition, formulation, interpretation, and validation. Brody suggests that the concept of the "life-span narrative" can assist in understanding moral problems through story (*Stories* 143–170). Hunter submits that narrative furnishes bioethics "a concern with time and change and causality" ("Overview"). Anne Hunsaker Hawkins argues that ethics cases—like other forms of literature—include moments of "epiphanic knowing." Throughout all of these literature-and-medicine functions primarily as a mere aid to bioethics, a handmaiden to the work of the philosopher. And, aside from a general criticism of the "top-down," mechanical nature of principlism, the relationship between the two disciplines has been generally friendly with the literature-and-medicine scholars assuring the ethicists that they do not wish to abandon normative judgments, ignore the moral principles, or fall into moral relativism. I contend, however, that literary theory should not simply assist ethics but should critique some of its central philosophic features.

CASES AS DATA

A concern with the nature of narrative should be of pivotal concern to bioethics because the ethics case is central to the discipline. The ethics case story had become, even before the revival of casuistry and the appearance of narrative ethics, the key genre of bioethics. In the present state of the art, bioethics, like medical discourse itself (see Hunter, *Doctors'*), is entrenched in stories. Ethical case narratives with commentaries are regularly published in national journals, such as the *Hastings Center Report* and the

Journal of Clinical Ethics, as well as in many small biomedical ethics newsletters. And, like members of other communities, medical ethicists speak to each other through shorthand references to a shared narrative folklore; ethicists do not have to retell an entire story but can simply say "Dax," "Helga Wanglie," or "Debbie" (Brody, *Stories* 144). One cannot sit through an extended discussion with medical ethicists without hearing cases presented to test some abstract argument. It is this movement from case to theory and then back again that gives bioethics its distinctive character. Robert Veatch summarizes this view: "General ethical rules are widely viewed [in medical ethics] as nothing more than rules of thumb, guidelines in clinical decision making that should direct its focus to each particular case. That the case is crucial to medical ethics is beyond dispute" (12).

Stephen Toulmin, in his often cited article "How Medicine Saved the Life of Ethics," explicitly associates the rise of bioethics with the "importance of cases." Toulmin contends that the moral issues in medicine beginning in the 1960s forced moral philosophers "to go beyond the discussion of general principles and rules to a more scrupulous analysis of the particular kinds of 'cases' in which they find their application" (737). This view is also presented in K. Danner Clouser's "Bioethics and Philosophy," which appeared in a special supplement to the *Hastings Center Report* on the history of bioethics. Clouser's article is a reflection on the relationship of philosophy to ethical questions that arise in medicine, but in his discussion of his work within the field for twenty-five years, Clouser brings up the issue of cases. Like Toulmin, Clouser contends that the "heart of the contribution" that bioethics has made to moral philosophy has been the "hard reality of its cases" (S11).

A good example of the case as data can be found in

an essay by James Childress introducing the "normative principles of medical ethics." Childress, at the outset of the essay, presents a case that "focuses on some major issues of moral justification and ethical reflection" (29–30). The case concerns a father who asks a physician to tell his family that he cannot donate one of his kidneys to his ill daughter for medical reasons although he in fact is histocompatible. After demonstrating how moral principles and rules can be used to resolve this dilemma for the physician, Childress offers some alternative theories that are critical of principlism. Childress explains, for example, that virtue ethicists have argued against principles and rules because they "emphasize (1) that a virtuous professional can *discern* the right course of action in the situation without reliance on principles and rules, and/or (2) that a virtuous person will *desire* to do what is right and avoid what is wrong" (29–30). Childress counters this argument by analyzing it in relation to his case: "When agents, such as the physician in Case 1, have to justify their conduct, it is not sufficient for them to appeal to their discernment or prudence or conscience without reference to principles and rules. There is simply no assurance that good people will discern what is right" (45). Childress then turns to the ideas of care ethicists, who argue that rules and principles are expressions of a male way of analyzing moral problems. Women "see and resolve moral problems differently," for they draw upon "narrative, context, and relationships rather than on tiers of moral principles and rules with a logic of hierarchical justification" as men do (45). Once again Childress finds an alternative to principlism wanting, for he asks "Would female and male physicians have viewed the problem differently in Case 1? Would they come to different conclusions? It is hard to predict with confidence that they would have ... because the major question is the *moral significance* of the nephrologist's relationships with the father and the other members of the family" (45–46). Childress criticizes these methods for not providing as satisfying a way to resolve his case. Indirectly he demonstrates how vital cases are to the way bioethicists think through moral issues. Regardless of whether the moral theory is based on deductive models such as principlism or inductive models such as casuistry, the case remains the data upon which one "tests" the theory. A theoretical approach that does not provide a way to resolve cases is deemed not only irrelevant but deceptive.

WHY THERE ARE NO ARTLESS CASES

In his article, Clouser argues that what truly differentiates the ethicist from the moral philosopher is not so much the use of the case as it is the use of "real" cases. "Trying out one's theory on real situations, thick with details, is very different from the philosopher's typical hypothetical case, which, if not simply invented, is so highly abstracted from real circumstances that only enough details remain to defend selectively the particular point the philosopher wants to make thereby" (S11). The difference, Clouser contends, has more to do with whether the case is presented to "illustrate" the theory (as it is for the philosopher) or is presented to "test" the theory (as it is for the ethicist).

This distinction between the philosopher's hypothetical case and the ethicist's real case has been continually reaffirmed by scholars who analyze how cases should be used in the bioethics discipline. Dena Davis, for instance, acknowledges that fiction can provide a useful source for studying ethical problems, but she maintains that the "daily bread of bioethics" is the "real" case. Furthermore she insists that these real cases keep the bioethicist honest, for "by describing real experiences ethicists can make points and draw conclusions while inviting their readers to make their own independent judgments" (13). Terry and Williams state bluntly that bioethics, unlike literary studies, "is about real people with characters, in contexts, during real time. Case studies miss this; hypotheticals destroy it" (19). Similarly John Arras, in his discussion of the pedagogical value of casuistry, counsels against using fabricated cases

> because hypothetical cases, so beloved of academic philosophers, tend to be theory-driven; that is, they are usually designed to advance some explicitly theoretical point. Real cases, on the other hand, are more likely to display the sort of moral complexity and untidiness that demand the (non-deductive) weighing and balancing of competing moral considerations and the casuistical virtues of discernment and practical judgment (*phronesis*). (37)

William Donnelly also cautions against using the hypothetical case, for "Such histories are usually constructed to illustrate the application of theory to concrete situation. The plot and characters are begotten of theory, not of life, and exist to demonstrate and confirm theory" (10). For these ethicists, hypothetical cases are biased, theory-driven, and constructed. Real cases are by implication impartial, theory-free, and guileless. The danger of "made up" cases, they suggest, resides in the teller's intentions to illustrate a prior theory; real cases, because of their origin in actual events, can question rather than support a philosopher's moral analysis.

This concern is continually reinforced by ethicists' tendencies to reassure readers that the cases presented are "real" or "actual." Tom Beauchamp and Laurence McCullough, in the preface to *Medical Ethics: The Moral Responsibilities of Physicians*, state that each of the cases they discuss "is based on actual events" (xv). In *Cases in Bioethics*, Carol Levine and Veatch note in their introduction that all the cases presented "are based on real events" (x). And in the acknowledgments to *Mortal Choices*, Ruth Macklin mentions that "all material is taken from actual cases" (ix).

Gérard Genette refers to assertions like this as forms of paratexts, which function "to ensure that the text is read properly" (*Paratext* 197). Furthermore Genette notes that these paratexts entail a "contract of truthfulness," which can be contrasted to a "contract of fiction" in which the author asserts that events or characters have no basis in reality. These contracts of truthfulness for a made-for-television movie, for instance, are supposed to increase the interest on the part of the viewer, a voyeuristic excitement that does not exist for a text that does not have to make this claim of truthfulness.

Contracts of truthfulness in bioethics texts exist, I suspect, to inform readers that the cases presented will test the validity of their theoretical analysis. But these assurances raise questions themselves concerning whether the reader would be able to determine the difference between the fictional case and the real case. Without these paratexts how does the reader know that the ethicist is truly testing the moral theory? Clouser seems to believe that the difference in the telling of "real" situation lies in its being "thick with details," the "relentlessness of the details" of the real case is what separates the moral philosopher's work from the ethicist's. Roland Barthes has noted that for readers the accumulation of details signifies a text as "realistic" ("Reality"); in other words, what Clouser sees as the mark of the real case is itself simply a representational effect that tells a reader that a text is not pure fiction. Without these markers—Genette's paratext and Barthes's reality effect—the reader of an ethics case would not be able to determine whether a particular text is fictional or not.

But regardless of the real status of a case's reality,

the reason that ethicists proclaim the superiority of the real case in bioethics has to do with the belief that unlike the hypothetical case, the real one is not constructed to provide a particular moral viewpoint. Bioethics is thought unique not because its cases can be shown to be objectively real but because as a matter of faith the real case will test rather than illustrate a moral theory. If there is any strongly held article of faith within the discipline, it is that bioethicists deal with the Aristotelian messy "real world" and that academic philosophers spend their time in a Platonic domain of unclouded abstraction. Bioethicists confront actual cases; academic philosophers contemplate imagined ones. A return to Childress's analysis above illustrates how this binary split between the ethicist and the moral philosopher is in actuality a false dichotomy.

Childress's criticism of virtue and care ethics is based on a naive—and common—misunderstanding of exactly what he is testing these two alternative ethics approaches against. His criticism suggests that he believes he is testing the approaches against stable facts, but this stance does not acknowledge that these cases are not themselves "the moral world" but rather *representations* of that world. Childress is not alone in this misunderstanding. A famous painting by René Magritte titled "La Condition Humaine" embodies this problem quite well. In the painting, we see a window open to a landscape but directly in front of the window is a painting which depicts the landscape so accurately that if the wooden legs of the easel and the side of the canvas were not visible, we would not be able to tell the difference between the two. We are so drawn into the painting that it is easy to forget that we are actually looking at a *painting* of a painting. Magritte's work demonstrates how we can easily confuse our representations for reality, as well as the impossibility of getting behind how we represent the world to see the world as it "truly" is. Similarly, ethicists have been testing and fine-tuning their theories on "realistic" representations of moral problems but do not acknowledge that there is a difference between the painting and the landscape (even one constructed to appear "realistic"). This difference lies in how any form of representation (visual or verbal) uses conventions to signify the world. As W. J. T. Mitchell observes, "Every representation exacts some cost, in the form of lost immediacy, presence, or truth, in the form of a gap between intention and realization, original and copy" (21). What ethicists have generally ignored is that cases—the data by which they test the relevance of moral theory—are fictions. That is, they are made up, constructed and thus follow conventions of representation that inevitably bias how one understands this information. Literary theory, therefore, is not simply a helpful assistant to bioethics but actually provides vital information and criticism concerning the fictional properties of the discipline's data.

A return to Childress's criticism of care and virtue ethics reveals how vital narrative theory can be for bioethics. Here is the case narrative in full that Childress uses in his analysis, a case presentation that is quite typical in bioethics for the representational conventions upon which it draws.

> For the last three years a five-year-old girl has suffered from progressive renal failure as a result of glomerulonephritis. She was not doing well on chronic renal dialysis, and the staff proposed transplantation after determining that there was a "clear possibility" that a transplanted kidney would not undergo the same disease process. The parents accepted this proposal. It was clear from tissue typing that the patient would be difficult to match. Her two siblings, ages two and four, were too young to be organ donors, and her mother was not histocompatible, but her father was quite compatible. When the nephrologist met with the father and informed him of the test results, as well as the uncertain prognosis for his daughter even with a kidney transplant,

the father decided not to donate one of his kidneys to his daughter. He gave several reasons for his decision: In addition to the uncertain prognosis for his daughter, there was a possibility of a cadaver kidney, his daughter had already undergone a great deal of suffering and he lacked the courage to make the donation. However, the father was afraid that if the family knew the truth, they would blame him for allowing his daughter to die and then the family itself would be wrecked. Therefore, he asked the physician to tell the members of the family that he was not histocompatible, when in fact he was. The physician did not feel comfortable about carrying out this request, but he finally agreed to tell the man's wife that the father could not donate a kidney "for medical reasons."

Childress has taken this case from an article on teaching ethics in the journal *Pediatrics* (Levine, Scott and Curran). If we put aside the issue raised by the content of the case and instead attend to the form of the presentation—that is, to its narrative qualities—we can see that part of the reason that Childress finds virtue and care ethics ineffectual in understanding this case has less to do with the theoretical limitations of these approaches than with the limitations of Childress's narrative form. Let us look at one feature of the narrative discourse—point of view—in Childress's case.

Narrative theorists argue that narratives can be categorized by the way consciousness is presented in a story. As Wallace Martin summarizes, "Access to consciousness' has two meanings: a third-person narrator can look *into* a character's mind or look *through* it. In the first case, the narrator is the perceiver and the character's mind is perceived. In the second, the character is the perceiver and the world is perceived" (143). To look into a character's mind is to treat consciousness as an object to view; to look through the character's mind is to treat consciousness as a lens with which to view objects. In Childress's story, the narrator looks into the consciousness of the characters: For example, "the father was afraid that if the family knew the truth, they would blame him," and "The physician did not feel comfortable about carrying out this request." Although the narrator tells us the thoughts and feelings of the characters, the narrator does not present events through any of the characters' minds.

The reason for this choice of point-of-view might be thought to be the result of presenting a third-person narrator. A narrative does not, though, have to be told in the first-person in order to view events through the consciousness of a character. The narrator of this case could have told the story through the consciousness of any (or all) of the characters without having to resort to the first-person. Although there are further distinctions that can be made about point-of-view within this case narrative, I simply wish to comment here that the way the case is written has implications for analyzing the moral problem, for if we return to Childress's critiques of virtue and care ethics, we see that it relies substantially on the concept of "perception." Childress states that there is no way to assure that the virtuous person will be able to "discern" the proper action, and asks if female and male physicians would have "viewed" the case differently. In both instances, Childress questions how we could know how the actors perceive the events. How would we tell if there is a difference between the perception of the virtuous versus the non-virtuous? How would we tell if there is a difference between the male and the female perspective on events? Yet I contend Childress's problem has more to do with the limitations of the data he has selected than the theories, and these limitations have to do with how consciousness is presented in the case. The narrator of the case looks into the minds of the characters in the drama, but does not narrate the events to the reader through the consciousness of any of the characters. Lacking this information, virtue and care ethics appear irrelevant because they address issues not present in the case. The problem lies not in what they address but in the fact that the case (like most cases in bioethics) does not provide the kind of information that makes its critiques relevant, namely, events as perceived by the various agents. A concern with the narrative conventions of ethics cases is thus central to understanding how Childress, like other bioethicists, tests the theories that challenge his own theoretical position. Although I do not charge Childress with intending to confound these theoretical positions in his choice of case (and thus his choice of form), I do believe he has drawn upon a narrative convention that rhetorically supports his own theoretical position.

It is because cases are pivotal to the task of bioethics and because cases are a narrative genre that I argue that the tools of narrative theory are central to revealing the rhetoric of the discipline. This study looks more deeply into point-of-view and at other narrative conventions (such as characterization, plot, authorship, and reading position) that are significant to analysis of the ethics case and explains how these conventions affect the reader's perception of the relevant features of an ethics case.

INTERTEXTUALITY AND THE CASE

Before exploring the fictional—or made-up qualities—of the bioethics case, we must first become cognizant of how the genre of the ethics case is intimately related to the medical case narrative. Julia Kristeva observes that all texts are comprehensible only because we see them in relation to other texts, for "in the space of a given text, several utterances, taken from other texts, intersect and neutralize one another" (36). It is because of these various other texts, or *intertexts*, that we have expectations for another text, and previous encounter with these other texts makes it possible for us to interpret the present text. This is not simply a matter of making allusions to other texts but rather a more important property by which the present text makes the reader aware of previous texts and then either affirms or subverts these expectations. Compare reading Zane Grey's *Riders of the Purple Sage*, listening to a radio show of *The Lone Ranger*, watching Clint Eastwood's *Unforgiven*, and reading Jonathan Lethem's *Girl in Landscape*. All four of these narratives achieve their effects through the expectations we have for the genre of the American Western. Grey's novel and the radio show fulfill our expectations, Eastwood's film subverts them for moral effect, and Lethem playfully combines them with the conventions of science fiction. In each instance, our understanding of and reaction to the texts depends on our understanding of previous texts. These tend not to be a single text but an entire web of texts that form our reaction. These four texts also depend on the various intertexts of their various mediums. For example, a listener to the radio play has expectations about the conventions established within radio programs. Consequently, in order for any text to be meaningful it must rely upon so many intertexts that are not all explicitly traceable. As Barthes observes, "The intertextual in which every text is held ... is not be confused with some origin of the text: to try to find the 'sources,' the 'influences' of a work, is to fall in which the myth of filiation; the citations which go to make up a text are anonymous, untraceable, and yet *already read*: they are quotations without inverted commas" ("Work" 160).

The medical case narrative serves as a pivotal intertext to most medical ethics cases. The bioethics case relies upon the medical case for much of its meaning but diverts from the medical case in pivotal ways. This will be perhaps most clearly seen in the issue of "reportability," which concerns what makes something "worth narrating," but the bioethics case narrative depends upon the medical case for many of its structural features. Contemporary concerns with questions in medical ethics arise in part as a public belief that the medical ethics case has become a different story requiring different readers from the medical profession. It is for this reason ethics cases often seem to read as if they were medical cases and then suddenly are transformed into something else. As I will argue, this deviation is expected and if it did not happen we would be equally disappointed. The television series *The X-Files* draws upon intertextuality to achieve its dramatic effects. The cases are assumed to be crimes that would normally be a concern for the FBI but turn into "ghost stories." Without this deviation, the fans of the show, I suspect, would be disappointed, but it is important to note that the series depends upon the expectations of the standard text of an FBI investigation. Similarly the ethics case always exists in relation to the medical case history.

Another important intertext for bioethics is the so-called hypothetical case of the moral philosopher mentioned already. Yet the tendency for ethicists to make comments in their prefaces that the cases they present are "real" or "actual" subtly subverts their assertions that there is any real difference between these types of cases for the reader. If the reader cannot determine the hypothetical from the real by reading the case then we must wonder 1) why this difference matters and 2) whether this difference exists. The first question, as I have noted already, has to do with the importance of separating the ethicists from the moral philosopher; in short, it has its roots in a disciplinary dispute within academia and matters to bioethicists, many of whom were educated as traditional philosophers.

The second question—whether there *is* a difference—is more serious, for it questions whether this dispute is actually a false dichotomy. It is perhaps better to replace the idea of the "real" with "verisimilitude." In narrative theory, verisimilitude refers to the text's ability to conform to the norms of a particular social world. These norms appear so natural to us that we do not even view them as "norms" but rather as simply part of what distinguishes the "real" world from the fantastic. For example, the genre "magical realism" only has meaning if we do not accept magic as conforming to these norms, yet, in another culture, magic in a story could be taken as the norm of that world and thus the concept of magical realism would lose its meaning. Tzvetan Todorov makes a similar point in his definition of the "fantastic" as a genre; he argues that the reader is unable to determine if the events are natural or supernatural (*The Fantastic*). Bioethics cases thus must possess verisimilitude in order to be accepted as data. The bioethics cases can be contrasted with such "fantastical" cases as Judith Jarvis Thomson's renowned "violinist case":

> You wake up in the morning and find yourself back to back in bed with an unconscious violinist. A famous unconscious violinist. He has been found to have a fatal kidney ailment, and the Society of Music Lovers has canvassed all the available medical records and found that you alone have the right blood type to help. They have therefore kidnapped you, and last night the violinist's circulatory system was plugged into yours, so that your kidneys can be used to extract poisons from his blood as well as your own. The director of the hospital tells you, "Look, we're sorry the Society of Music Lovers did this to you—we would never have permitted it if we had known. But still, they did it, and the violinist now is plugged into you. To unplug you would be to kill him. But never mind, it's only for nine months. By then he will have recovered from his ailment, and can safely be unplugged from you." (154–55)

I believe that this is the kind of case that ethicists tend to discount as unworthy data for it seems so overtly constructed to simply support the moral position in Thomson's argument. But the problem is not so much that this case is hypothetical but rather the reader knows (or perhaps one should say believes strongly) that this is not a "real" case because it goes against the conventions of verisimilitude for ethics cases. This type of case is still an important intertext as the kind of cases that are thought to be implausible and thereby not worth serious attention within the discipline.

FRAMING THE CASE

Narrative theorists like sociologists, art critics, and linguists have been highly cognizant of the importance of framing in social life. Framing encloses something and thereby sets it off from other forms of communication and interaction. Frames act as signals that what is inside should be attended to differently from everything else. Magritte's work noted above explicitly plays with the importance of framing to differentiate representations from the "real" world. Although the framing of a piece of art is an obvious example of the power of this concept (and the work of Marcel Duchamp demonstrates how powerful frames can be to "creating" art), framing has come to be used as a general term for the action of setting things apart in order to indicate that the framed entity should be treated separately from other forms of discourse. Gregory Bateson observed that animals frame certain activity as "play" in order to differentiate real (and thus dangerous) aggression from pretend aggression. Erving Goffman was particularly attentive to how social behavior is framed to present the self in an ideal way and to hide those features that may cause doubt to this idealized self. Framing in both art and social interaction indicate to observers that this is discourse worth focusing on and this discourse should be treated differently ("this is art").

Literary discourse also has ways of framing. Some of this framing occurs through the physical act of the print medium. Classics are deemed worthy of being bound in leather; romances only in cheap paperbacks. Articles that are printed in respected journals receive different attention from articles that have been hand-written on yellow legal pads; the same words appear but our attitude toward them is qualitatively different. Genette points out that "paratexts," which include such items as titles, prefaces, designation of authorship, notes, and dedications help frame the particular discourse. How would a potential reader know that a work is fiction and not non-fiction if the publisher did not give this information in the form of labeling on the book's cover? The preface in bioethics has been an important paratext to inform readers of the "truth" of the cases presented. Obviously philosophers are concerned that without such paratexts the reader would not be able to understand the importance of these cases.

Loretta Kopelman's article "Moral Problems in Psychiatry" in the anthology *Medical Ethics* includes thirteen cases. Each of the cases is separated from the rest of the text by being placed in a "black box." Throughout the article, Kopelman's cases are physically presented as separate from her philosophical analysis. This separation reinforces the illusion that the case presentation is not tainted by the philosophy. The creation of this block is only one way in which bioethicists and their editors have framed the case. Sometimes cases are indented, italicized, or printed in a different type. Cases can also be framed by being collected in an appendix. In each instance the act of framing presents the case as separate data and not in any way influenced by the philosophy. This act of framing is important in order to maintain the fact/value distinction that has been important to the field. If cases were not framed from the philosophy then bioethicists could not argue that their work is any different from the hypothetical thought-experiments of the moral philosopher. In this book, as we will see, I wish to suggest the frame is very much a fiction.

What follows is a series of "readings" of ethics cases. I argue that ethicists have written cases in a manner that supports their philosophical positions, but I do not see this work as primarily an "exposé" of the discipline's terrible secrets. Instead, through these readings, I hope to provide a model for a self-reflexive bioethics. A similar self-reflexive posture has been developed in other academic disciplines. What many of these disciplines have in common is that their data is a literary construction. Consequently, anthropologists (such as Clifford Geertz), historians (such as Hayden White), economists (such as Donald McCloskey), and philosophers of science (such as Thomas Kuhn) have argued that attention to the historical, social, and rhetorical constructions of disciplines is a necessary move in academic honesty. Recognizing that the data of one's discipline is a fiction, made up, should not result in an abandonment of the discipline but rather a desire for as much rigor as possible in the analysis of that data. Because of this desire, I have turned to the tools of narrative theory; I believe that the best way to read the data of bioethics is through the tools of what they are: that is, narrative. In each of the chapters of this book, I wish to move the discipline of bioethics toward such a rigorous criticism in the hope that we become conscious of what we have been applying our moral theories to.

Chapter Two: From the Ethicist's Point of View

"Every way of seeing is also a way of not seeing."
>   —*Kenneth Burke*

Ethical analysis ought to concentrate as much on how one sees moral dilemmas as deciding how one should act in response to them. As Stanley Hauerwas observed, "Modern moral philosophers have failed to understand that moral behavior is an affair not primarily of choice but of vision" (34). And William May has resounded, "Moral reflection attempts, at its best, a knowledgeable revisioning of the world that human practice presents.... We cannot change our behavior unless, in some respects, our perception of the world also changes" (15). Yet, in responding to issues in medicine, the ethicist who is arguing for a new vision is also the author of the old vision, which they have presented in the form of an ethics case story. Stories of moral dilemmas, like all narratives, are constructed from a specific point of view, which persuades us to see the events in a particular manner. In this last statement, critics might claim that I am confusing the literary point of view with the philosophical point of view, confusing a literal concept of who sees with a metaphorical concept of what one believes. This curious interchangeability of words for physical perspective and philosophic perspective (such as position, slant, direction, stance), Susan Sniader Lanser argues is not accidental. She notes that "however reluctant the critical tradition has been to integrate questions of value with the analysis of form, literary scholars still understand that one's attitude has everything to do with where one stands, that technique is never wholly independent of ideology" (Lanser, *Narrative* 17–18). By situating the teller physically one is also able to situate the teller philosophically. In this chapter I wish to demonstrate how Lanser's insight applies to the ethics case.

What is often referred to as point of view, the literary critic Gérard Genette reminds us, encompasses both mood or perspective (the issue of who sees) and voice (the issue of who speaks) (*Narrative* 186). If bioethics is to make use of case narratives, it must begin to take seriously the point of view expressed in these accounts of moral dilemmas. Because one is asked to take a moral position in response to these stories, one must become cognizant of how the perspective and the voice of the discourse affects one's vision of the moral world. In order to explore the relation between moral and narrative points of view, I examine in this chapter three examples of literary point of view and how the viewpoints from which the tales are told act to persuade readers of particular assessments of ethical situations.

FROM THE CLINICIAN'S POINT OF VIEW

Terrence Ackerman and Carson Strong, in the preface to *A Casebook of Medical Ethics*, explicitly state that almost all the cases in their textbook are derived from their experience in the clinical setting and are "accurate accounts of actual cases." All of the cases, save three, "were typically encountered during clinical rounds or special consultations" and were "discussed extensively" with the health-care team. They also note, "In many cases we reviewed aspects of the situation with patients, family members or other significant participants" (viii). These assertions suggest their interest in establishing a personal relationship to the cases they are telling in much the same way Clifford Geertz argues that ethnographers have traditionally striven to establish authorial presence, or what he terms "signature" (9). According to Geertz, ethnographers wish their readers to know in no uncertain terms that they were "there," and by establishing their signature to the descriptions, they convince their readers of the accuracy of these accounts of strange worlds. In *A Casebook of Medical Ethics*, Ackerman and Strong likewise establish their signature and through it their authority for the telling. They, like ethnographers, wish to demonstrate that they were "there." Unlike many collections of bioethics cases that lack this establishment of "signature," Ackerman and Strong assure us, in effect, that the cases have no invisible quotation marks around them—that is, they are not merely quoting someone else's account of events. In this their text is distinctive. The reader can, thereby, *situate* these particular ethicists in relation to the events narrated in these cases. In responding to an early reviewer's comments implying that the cases were in some way "altered or fabricated," Ackerman and Strong assure their readers that they are attempting to present accurate portrayals of events. It is only in expanding the views of the various health care professionals "to identify more completely the relevant ethical views and considerations" that the authors have changed the cases to fit their philosophical concerns (ix).

Ackerman and Strong maintain a consistent narrative form throughout their textbook (the notable exception being chapter four on medical research, which deals with larger policy issues), and beyond a richness of medical and psychosocial details, their style of presentation is similar to many case presentations in medical ethics. Look at the beginning of the first case in their collection.

> M. J., a sixty-year-old man, was admitted to the psychiatric ward of the Veterans Administration hospital after he threatened to kill himself and his wife with a hunting rifle. The incident followed almost two years of increasing physical and mental difficulties. The patient had suffered continually from depression and often contemplated suicide. He admitted to sleep disturbance (early-morning awakening), loss of interest in outside activities, absence of sexual interest, and problems with concentration and memory. He also had a variety of nonspecific physical complaints (such as "weakness in the legs") and considerable loss of appetite.
>
> Formerly, the patient had been happily married for thirty-five years. He also had a good relationship with his only child, a thirty-three-year-old son who lived in the same town. He reported no special problems in childhood or adolescence and has never had a problem with alcohol or drugs....
>
> The vocational history given by M. J. was unremarkable. He worked for fifteen years as a salesman and during the last twenty-one years had been an auto body repairman....
>
> The patient was diagnosed as having endogenous depression. The term refers to depressive illness that is not a reaction to environmental stress (such as the death of a loved one), the implication being that it results from some intrinsic biological process.... (3–4)

After explaining Electroconvulsive therapy (ECT) and its clinical effectiveness, Ackerman and Strong continue their description of the situation.

> The problem in this case was M. J.'s ambivalence toward ECT. Several times he agreed to undergo ECT but then refused before therapy could be undertaken. Twice a series was initiated but stopped on his insistence.... Over several

weeks in which these futile attempts to complete ECT were occurring, the patient became more reclusive, was refusing to eat, and was exhibiting exacerbated depressive symptoms and bodily complaints.

The opening of this ethics case story is written in the style of medical case histories as physicians, residents, and students present them at morning reports and grand rounds presentations and in patient discharge summaries. Ackerman and Strong provide more background information and explanation of clinical definitions and procedures than a clinical presentation would. However, their case presentation, on the whole, appropriates many of the defining traits of medical storytelling: plot, passive constructions, and clinical linguistic features (concepts I am deriving from Anspach; Hunter *Doctors'*; Mintz). Note how Ackerman and Strong plot this case by beginning with M. J.'s "presentation" to the psychiatric ward and, following a description of his "present complaints," they move into the past to describe what led up to the current condition: "Formerly, the patient had been happily married for thirty-five years." For the physician the plot of the story is determined by diagnostic concerns (Hunter *Doctors'*), and Ackerman and Strong take their plot structure for this ethics case from medicine. They do not tell the patient's story, nor do they tell their story, that is, the ethicist's story; instead, they tell the physician's story. Yet one can assume if Ackerman and Strong related the narrative in terms of how *they* encountered M. J. or M. J.'s own experience it would have an entirely different plot.

Many of the linguistic features in Ackerman and Strong's text are intelligible only within the context of clinical medicine. The narrative can "make sense" only if one already shares with the narrator assumptions about how an ill person should be viewed. A story, like every other act of communication, is an act of collaboration, for information must be shared between the teller and the audience. Without these shared cultural norms, the narrative would be incomprehensible. The first sentence, "M. J., a sixty-year-old man, was admitted to the psychiatric ward of the Veterans Administration hospital after he threatened to kill himself and his wife with a hunting rifle" identifies this man through the clinical gaze. First it identifies him by his initials (rather than by name or pseudonym). The reader is also told M. J.'s age and sex, and then the verb "was admitted" indicates the first action performed on this individual. Unlike writers who wish to depict a character who has some event occur to him or her, Ackerman and Strong "present" a patient. They also use the contrasting signs of "admission/denial," a characteristic binary split within medical discourse: "He admitted to sleep disturbance" and is said to have "reported no special problems in childhood." All of these linguistic features are borrowed from the way a clinician views a patient (Anspach; Mintz).

When describing actions taken upon this patient, Ackerman and Strong primarily use the passive voice. When physicians present patients to one another the passive voice becomes a covert code of insidership, of a shared viewpoint. The patient is the one acted upon, but the subject of the sentence is an implied, and sometimes explicit, "we." Yet this is an ethics case, not a medical case; its teller is not a physician but two ethicists. As a result the use of passive construction acts as a secondary sign, communicating that the ethicist is one of the implied agents. Passive constructions here translate into "we, the physicians and the ethicists, admitted M. J. to the psychiatric ward." In each of the three aspects of their narration—plot, language, and passive verb constructions—Ackerman and Strong adopt the clinician's voice and thereby the clinician's authority. They are not quoting a case presentation but, in effect writing it themselves, assuming a clinician's presentational style and particular viewpoint in telling about an ethical problem.

After describing the problem, Ackerman and Strong move away from the passive voice and thus apparently from the clinician's perspective. At this point in their narrative, the psychiatrist becomes an "actor" in the drama. When describing M. J.'s refusal of the recommended treatment, Ackerman and Strong depart from the plot, language, and passive construction borrowed from the physician's collective first-person case history and adopt the perspective of a third-person narrator. This shift in narrative discourse occurs repeatedly in Ackerman and Strong's cases as the writers change over from a first- to a third-person narrator who, because they are situated outside the quandary, leads the reader to believe that what is being provided is an apparently uninvolved view. This change suggests that there has been a shift from a personal narration, a story told from the viewpoint of a particular character, to an apersonal narration, a story told by an impartial observer. It is important to note that the passive voice, which hides the agent of action and is so much a part of the standard form of the medical case history, exists primarily to suggest that the narrative is apersonal, that is, the narrative is objective and "scientific" (cf. Anspach). Previous to this shift, the authors employ the covert first-person plural of the medical case history, and the reader assumes that they depart from the perspective of a physician as they use the third person.

Do Ackerman and Strong truly take an impersonal stance merely because they have begun to write explicitly in the third person? Not necessarily. Roland Barthes has argued that some narratives are in actuality not apersonal but covert forms of *personal* ones. He observes that, "there are narratives or at least narrative episodes ... which though written in the third person nevertheless have as their true instance the first person" (*Image* 112). Barthes asserts that the covert personal can be distinguished from the truly apersonal by rewriting the narrative and changing the pronouns from *he* or *she* to *I* "so long as the rewriting entails no alteration of the discourse other than this change of the grammatical pronouns, we can be sure that we are dealing with a personal system" (112). One can in turn determine whose narrative it is; one can situate the narrator.

Here is a section from Ackerman and Strong's case narration:

> The attending psychiatrist envisioned three options: (1) seek to have the patient declared incompetent to make treatment decisions; (2) threaten him with involuntary commitment to a state hospital unless he accepted ECT; or (3) continue to review the potential benefits and risks of ECT with the patient.
>
> Each option had its difficulties. To begin with, the psychiatrist was not convinced that M. J. *was* incompetent. Several lengthy discussions about ECT with the patient failed to yield clear and recurring reasons why he refused treatment, although he once mentioned a fear that ECT might kill him and that it was causing his eyesight to deteriorate.... He also had very poor insight into the seriousness of his condition. (5)

If one assumes this new narrative voice in M. J.'s case and substitutes the pronoun "I" for "the psychiatrist" one finds that the narrative coheres.

> I envisioned three options: (1) seek to have the patient declared incompetent to make treatment decisions; (2) threaten him with involuntary commitment to a state hospital

unless he accepted ECT; or (3) continue to review the potential benefits and minimal risks of ECT with the patient.

Each option had its difficulties. To begin with, I was not
convinced that M. J. *was* incompetent. Several lengthy discussions about ECT with the patient failed to yield clean
and recurring reasons why he refused treatment, although
he once mentioned a fear that ECT might kill him, and that
it was causing his eyesight to deteriorate.... He also had
very poor insight into the seriousness of his condition. (5)

If one rewrites the case a second time and instead substitutes "I" for the pronouns used for the patient, the
passage soon rings false. One could possibly accept the
statement "The attending psychiatrist envisioned three
options: (1) seek to have me declared incompetent to
make treatment decision....," but when one gets farther
in the narration the pronoun substitution does not
hold up:

On the other hand, the procedures, benefits, and risks of
ECT were explained on several occasions, and I seemed to
comprehend the information. This was suggested by my tendency to frequently consent to ECT before later withdrawing.

At this point in the narration the sentence makes
sense if a physician is relating the information, but M.
J. as the narrator seems implausible. It is absurd to
think that the patient would state that "I seemed to
comprehend the information," and "This was suggested by my tendency," for the words "seemed" and
"suggested" involve judgments consistent with an
observer who doubts. Nor is this the viewpoint of an
omniscient apersonal narrator. The perspective in
the narrative continues to be that of the physician and
that despite the move from first person to third person, there has not been a genuine shift into an apersonal narration.

That medical ethicists who are not trained clinically
often adopt a clinical persona in presenting cases has

rarely been noted, much less criticized. The point of view adopted in Ackerman and Strong's case presentations is one commonly found in bioethics, and it carries with it important messages about the discipline. In part, it persuades readers of the legitimacy of bioethics in the medical setting by portraying moral inquiry as possessing the same features as medical inquiry. Ackerman and Strong note that their own analysis of the issues will follow two approaches to evaluating moral actions. The first is the ranking of particular moral principles in a specific case, and they note that this has been the traditional way conflict of values have been resolved within bioethics. The second approach employed by Ackerman and Strong is the "balancing approach," which attempts to arrive at some form of compromise among principles to resolve problems. They contend that "a balancing approach as the resolution to the moral problems is more commonly utilized in the clinical practice of medicine. Physicians frequently resolve moral issues by utilizing policies that take partial account of each of the initially conflicting values or obligations" (ix). The authors argue that their "systematic development" of the balancing approach "represents an attempt to broaden the debate about how to resolve moral issues in clinical medicine" (xi). Hence Ackerman and Strong acknowledge that one of the unique features of their analysis is their systematic representation of the physician's point of view. The narrative perspective supports and anticipates the philosophical perspective.

Both approaches, however, follow the principle-based tradition within bioethics of determining conflicting principles within a case; they differ only in the manner in which a conflict should be resolved. Edmund Pellegrino, in his historical survey of bioethics, astutely remarks on the appeal of principle-based approaches to the medical community: "It … offered an orderly way to 'work up' an ethical problem, a way analogous to the clinical workup of a diagnostic or therapeutic problem" ("Metamorphosis" 1160). The clinical slant in ethics case-presentation has been the way many principle-based ethicists have customarily presented their problems, and Ackerman and Strong's style of presentation is remarkable only in the degree to which they reproduce the clinician's point of view. Interestingly, they add a final note that cautions readers not to valorize the physician's point of view: "the frequent use of this [balancing] approach in clinical practice is not sufficient to justify its use. Thus, the thoughtful reader is invited to assess the comparative merits and liabilities of these different conceptual frameworks for resolving the difficult moral dilemmas that confront the conscientious physician" (xi). Ackerman and Strong's work reflects the many tensions within medical ethics that come from being a part of medicine yet wishing to remain apart from it. The narrative point of view they adopt reflects the role ethicists have had within medicine, sometimes putting on white coats, acting as consultants, working up ethical problems, and writing chart notes with moral prescriptions. Their style of presentation nevertheless raises questions of the degree to which one can rhetorically appropriate the point of view of another while remaining critical of that other's perspective.

FROM THE OBSERVER'S POINT OF VIEW

Baruch Brody, in *Life and Death Decision Making*, supports his argument for a pluralistic approach to ethical dilemmas through an analysis of forty ethics cases. Like Ackerman and Strong, he establishes his signature to these cases in a preface. Brody contends that his cases "are composites drawn from several hundred real cases I have encountered in teaching rounds and/or consultations," and although none of these "real" cases "corresponds exactly" to the facts in his written cases, all the facts are "drawn from a real case." Moreover his accounts of the medical team's arguments "are drawn from arguments actually offered by sensitive and talented clinicians in several hundred real cases" (vi). Brody specifies that he was there in a professional capacity. Like Ackerman and Strong, he seems eager to establish the authenticity of the cases. He repeats the words "real case" four times in one paragraph. Brody also emphasizes that the opinions of clinicians he reports are derived from those "actually offered." Brody wishes us to know that he has been among the natives, and his descriptions are genuine and authentic because he can lend his signature to them. However, Brody does not, beyond reassuring the reader of the authenticity of the cases, speak of how he determined the style of presentation he uses in telling these cases. Furthermore, determining Brody's exact relationship to the participants in the cases is substantially more difficult than for Ackerman and Strong. His presentation of case eleven, for example, uses a point of view substantially different from Ackerman and Strong's. Brody separates the "facts" from the "questions," which as he notes in his preface are usually the "arguments presented by the team."

> FACTS Mrs. K is a 69-year-old woman diagnosed as having adenocarcinoma of the lungs. Surgery to remove the primary tumor was ruled out because of a local lymph node involvement. She received radiotherapy both for her lung disease and for more recent metastases to the brain. She then became blind. [sic] probably because of optic nerve compression as a result of her metastatic disease. Everyone is amazed that she is still alive, but no one believes that she has much longer to live. She is very depressed, more by

blindness than by her impending death, and she won't attempt to learn any skills. All she keeps asking is whether or not she can be operated on to remove the local compressing masses so that she can see again. Her husband supports this request. A social worker has spent a fair amount of time working with them, explaining that the surgeons don't wish to operate in light of her very short life expectancy and the uncertainty of success of this difficult surgery and that Mrs. K would do better to learn certain elementary skills so that she can make the best of the time left to her. She and her husband refuse to accept his idea. She says that she wants to see again. He says that all he cares about is that she can have her chance to see, and he is very angry at the surgeons for refusing to operate.

QUESTIONS ... Some people see Mrs. K as being very depressed by her blindness and impending death and insisting on surgery to correct her sight as a way to avoid dealing with the thought of dying blind. Others argue that there is insufficient evidence of depression to challenge the judgment of competency. They insist that those who challenge her competency are doing so simply as a way of not agreeing to her wishes.... Everyone has a great deal of pity and compassion for this woman, who is dying blind, and her loving husband, who wants at least some of her wishes to be fulfilled. (136)

Although in the preface Brody acknowledges that he was present in these cases, that like Ackerman and Strong he was "there," the reader has difficulty situating him in the flow of the events. The first few sentences suggest that he has also adopted the clinician's viewpoint, "Mrs. K is a 69-year-old woman diagnosed as having adenocarcinoma of the lungs," and afterward he maintains many of the traits of medical discourse as well as its plot structure. In the next two sentences Brody uses the passive voice, "Surgery ... was ruled out," and "She received radiotherapy." Then Brody makes an observation that would ordinarily be out of place in a clinical case presentation: "Everyone is amazed that she is still alive, but no one believes that she has much longer to live." One cannot simply change the pronouns in Brody's narrative and expose a particular covert personal narrator. Instead these are truly apersonal, third-person narrations. Although the reader knows that Brody was involved in the case (or a case very much like it), in his telling of the events, Brody is an invisible observer, a secret sharer, who gathers all the "facts" and listens to all the voices. His position and participation in the events seems not to affect his description. He hears not only the clinician's perspective but also "observes" the social worker spending time with the family "working with them, explaining that the surgeons don't wish to operate." And finally Brody records, "She says that she wants to see again. He says that all he cares about is that she can have her chance to see, and he is very angry at the surgeons for refusing to operate." This case is not told from the clinician's point of view, nor the social worker's, nor the patient's. Even with Brody's explanation that this is a composite of several cases, it must strike the reader as odd that Brody has chosen not to place himself in the case.

The viewpoint of an unseen or nonparticipatory observer is maintained in Brody's summary of the discussion by the medical team. "Some people see Mrs. K as being very depressed by her blindness and impending death and insisting on surgery to correct her sight as a way to avoid dealing with the thought of dying blind. Others argue that there is insufficient evidence of depression to challenge the judgment of competency." Brody seems to move like a spirit, hovering from conversation to conversation. Invisible and distant from the ongoing events, he not only reports what is being said, but also presents a third-person view that the reader cannot challenge because it comes from an apersonal narrator. Even if Brody is one of the "others" who takes a different position, he is still referring to himself in the third person. Take the statement, "Everyone has a great deal of pity and compassion for this woman." Does the "everyone" in this statement include Brody? He is "present" but only as a silent witness to the ongoing events. Though one could argue that Brody is simply recording the information as it was related to him, that he was not actually present during any of the events or conversations with the family, there is nothing in the text that supports this. Instead this third-person narrator hides how the information was obtained, and the reader cannot determine at what stage Brody became a participant in the case. Did he observe all the events? Was he part of an ethics consultation? Was he called in to lead a discussion of the case with the medical team, and is he just relating the information they gave to him? One can situate Brody in the events that he is telling only because he has previously established a signature that positions him in relationship to this case; one cannot place him in the actual flow of the narrative, only as an observer of the events.

Boris Uspensky has categorized this type of literary point of view as a "sequential survey" in which "the narrator's viewpoint moves sequentially from one character to another and from one detail to another, and the reader is given the task of piecing together the separate descriptions into one coherent picture" (60). Uspensky compares this style of viewpoint to a film presentation in which the reader receives a montage of scenes. Similarly, Brody's narration delivers fragments of different perspectives that the reader must patch together, and there is no single perspective that holds the narrative together.

Although it is difficult to determine the manner in which the narrator was able to gather his information, the story is clearly controlled through the narrator's voice. This is most clearly revealed in the indirect reports of the different assertions: Mrs. K "keeps asking," her husband "says," the social worker is "explaining," some people "argue" and "insist." The various participants do not "speak" for themselves directly but always indirectly through the narrator, that is, they are never quoted directly. Brody's viewpoint, though, is not that of an omniscient narrator but that of a limited one. Genette distinguishes between internal and external focalizations in narrative (*Narrative*). Internal focalizations are narrative told from the point of view of a character within the narrative; external focalizations are told by a narrator who is an external observer and is limited by an empirical perspective. Brody does not go inside the minds of the participants or present information to which some of the participants would not have access, such as the "true" motivations of one of the characters or the internal feelings or thoughts of the participants. In Genette's terms, Brody's is a narrative of external focalization. While in the preface Brody is eager to show that he was an insider, the narrative viewpoint he adopts belies this stance.

These stylistic features reflect Brody's interest in a pluralistic approach that "accepts the legitimacy of a wide variety of very different moral appeals" (9). Confronted with often equally compelling arguments, Brody argues against approaches that provide a hierarchy of values or that attempt to provide a contextual scale for moral conflicts, like the priority and balancing approaches discussed by Ackerman and Strong. Advocating a process of judgment in which "We look at the various appeals and their significance, and then we judge what we ought to do" (77), Brody acknowledges the inherent messiness in which this may leave moral decisions. He asserts, however, that bringing into focus the various conflicting values is "the most a moral theory can provide" (79). Brody thus narrates his ethics cases from the point of view of an impartial observer capable of judging the various appeals. This textual point of view, then, is consistent with the moral perspective that Brody advances. Although involved in the cases, Brody recalls the narrative through external focalization—that is, as an uninvolved observer. The reader of Brody's narrative sees the problem through the perspective of a supposedly unbiased judge who regards the various conflicting appeals. The viewpoint of the narrator is not natural as a personalized "I" would be, nor conventional like the effaced "we" of medical cases, but one that Brody has chosen and constructed and one that supports the way he wishes the reader to see and evaluate the moral questions.

FROM THE PROTAGONIST'S POINT OF VIEW

> I was alone, waiting for the start of an ethics case conference in an all-purpose room of the Child Development Center. I remember that the cement block walls of the 1960s vintage building were painted with a stark yellow (they must have wanted it to look like sunshine, to cheer up the handicapped, I said to myself). Feeling very much the displaced, developed adult consultant, I was in the grip of wondering whether the gray molded plastic chair would really continue to support me, when the pediatric resident, who would be presenting the case—and whom I will call Dr. McDonough—arrived and started telling me the following story while we waited for the other committee members.
>
> "The patient is 21 years old," he said, "is the size of a 7-year-old, and has the mental age of a 2- to 2-1/2-year-old." (Reich 279)

The opening of Warren Reich's "The Case of the White Oaks Boy" in his article "Caring for Life in the First of It: Moral Paradigms for Perinatal and Neonatal Ethics" is startling when one compares it with the cases presented thus far. Here is an instance of an ethicist who situates himself in relation to the events narrated, yet in Reich's tale the use of "I" has just as much rhetorical force as the narrative viewpoints of the clinician and the observer presented above. The first sentence radically locates the ethicist as the teller—"I was alone"—and the reader knows whose expression this is. An important issue concerning "authorship" should be noted, however. Reich's use of the first person leads the reader to assume that the author of the case and of the philosophical perspective are one and the same, and the reader also assumes that this establishment of signature through the first person indicates that this is a "real" case drawn from Reich's experiences. These are aspects of the literary conventions that readers bring to an ethics case presentation. Yet, as much work in literary criticism has observed, the "I" of a narrative (even a non-fictional one) need not be the same person as the author. It is the nonfictional nature of the story with the first-person narrative that leads the reader to assume that the two are one and the same. While the reader may infer this, Reich does not explicitly claim that he is the author of the case, and he treats this narrative in his analysis as if it were given to him by someone else. If Reich is the author, he never uses the pronoun "my" when referring to the case, and the distance he keeps between the philosopher and the storyteller suggests that he wishes to keep the two I's separate from one another. This is an issue that I will address directly in following chapters. The reader, however, is given clear signs in the story that, like Reich, the narrator is an ethicist, and I shall for this discussion refer to the narrator as "Reich" although I am aware that this is as much a construct as the other narrators in the cases already discussed.

The ethicist-narrator continues by recounting personal memories and even talking to himself, and he notes how he feels displaced. When the physician arrives on the scene, he breaks the ethicist-narrator's reminiscences with a conventional clinical presentation: "'The patient is 21 years old.'" (In Ackerman and Strong's style, this would be the first sentence of the ethics case presentation, and it would not have quotations marks around it.) The narrator's account begins not with a patient presenting himself to a physician but with a physician presenting a problem to an ethicist. The plot structure in this case is conditioned by the specific viewpoint of the ethicist-narrator. All the information that the reader receives about this problem is through the ethicist-narrator's perspective, and thereby all the information is attached to that particular voice. The physician tells the narrator of a patient whose quality of life they are unable to determine. Unlike Brody, Reich indicates that the ethicist-narrator plays a role in gathering information:

> Itchy to shift attention elsewhere, I asked my conversation partner: "Dr. McDonough, what was this patient like? How did he strike you? What did you think of him?" McDonough, his face now transformed by curiosity and amazement, told me what I (as a nonphysician) regard as the "real story" inside the case history.
> He said, "Michael is a very strange individual. He shows unusual behavior. I'll never forget him—how he seems to be capable of just three things."

In this, the ethicist-narrator maintains the status of an outsider or one who seems unconcerned with the details of clinical evaluation. The ethicist-narrator knows that there is a "real story" to be found, and it is not the case history but someone inside it. Finally the ethicist-narrator requests "spontaneously" to see Michael, "this boy-man," and as McDonough begins to examine his patient, the narrator describes an extraordinary moment of connection with Michael:

> By the time Dr. McDonough had raised his stethoscope to Michael's chest and touched him with it, I had already attempted to enter into Michael's mentality. I could sense something like a feeling of gratitude in Michael, reflected in his face as he stared at the device that was connected with McDonough's head: "Thank you, Dr. McDonough, for this beautiful tube of yours." Michael reached out and softly gripped the stethoscope as though it were part of his doctor-friend's body.
>
> I stood for a long time, never taking my eyes off Michael as long as he held that life-giving tube and stared at it with restful, smiling eyes.

The ethicist-narrator is made the protagonist of the narrative and tells the story from that viewpoint. This is not the narration by a minor character, (as for example the first-person narration by Nick Carraway in Fitzgerald's *The Great Gatsby*), for the ethicist is truly the focus of this narrative. It is not the ethicist-narrator's involvement in the events that forces this perspective, for the narrator could have portrayed the events from the perspective of an objective recorder. Instead, it is the ethicist's story. Compare, for example, this first-person narration with Rita Charon's in "The Case: A Relative Stranger" or with Timothy Quill's in "Death and Dignity: A Case of Individualized Decision Making." In these accounts, although Charon and Quill write in the first person, the patients are the protagonists, and the narrators remain secondary characters in the stories. Since I will discuss Charon's case in more detail in relation to gender issues, look at how Quill begins his narrative:

> Diane was feeling tired and had a rash. A common scenario, though there was something subliminally worrisome

that prompted me to check her blood. Her hematocrit was 22, and the white-cell count was 4.3 with some metamyelocytes and unusual white cells. I wanted it to be viral, trying to deny what was staring me in the face. Perhaps in a repeated count it would disappear. I called Diane and told her it might be more serious than I had initially thought.... When she pressed for the possibilities, I reluctantly opened the door to leukemia. Hearing the word seemed to make it exist. "Oh, shit?" she said. "Don't tell me that." Oh, shit! I thought, I wish I didn't have to.

Diane was no ordinary person (although no one I have ever come to know has been really ordinary). (691)

Although Quill has written this narrative through the point of view of a first-person narrator, he is involved but not the center of the narrative. His primary aim is to rhetorically convince the reader how "Diane taught me about the range of help I can provide if I know people well and if I allow them to say what they really want" (694). If Quill was the protagonist of the narrative, if the reader felt that he was driven by selfish desires, then readers may not conclude that he was a virtuous agent.

In contrast to Brody's sequential viewpoint, the unity of the "White Oaks Boy" narration is achieved through the protagonist's perspective—that is, the ethicist's point of view. The ethicist opens the case with an image of sitting alone and "displaced" and so the reader comes to believe that when Reich encounters Michael, who is also in many ways alone and "displaced," the narrator is able to establish a special connection. Just as the ethicist-narrator gets "inside" the case history in his conversation with the clinician to find the "real story," the narrator gets inside Michael to see the "real person." Is it surprising that the method that Reich advocates in understanding this case is an "experiential" ethics? According to Reich one should begin "with a perception and interpretation of values related to moral experience—that are conveyed through life experiences, narratives, images, models known from behavioral sciences, etc." (283). He proposes an ethics based on response to an "Other" rather than on abstract moral reasoning. By sensitively penetrating the inner world of patients, an ethicist, he believes, can determine how to respond to their needs. Reich's choice of a case that uses the first-person voice makes sense in view of his phenomenological orientation. It persuades the reader of the success of such an ethical position because he has provided an account that reveals a point of "experiential" epiphany.

In "Caring for Life in the First of It," Reich arrives at his support for an "experiential" approach after he has attempted to apply other, more traditional, paradigms in medical ethics. He is dissatisfied with each of these approaches for they do not truly provide aid in resolving the moral problem of his case. Tellingly, Reich comments that the "thrust" of his "experiential" method is "to break the preoccupation of ethics with reasoning stemming from the ethical analyst's point of view ... and recenter ethics on the stranger, by allowing his or her story to refocus our vision, and expose the relativity of our own orientation to what is meaningful" (285). Yet could the reader be persuaded of this argument against the previous paradigms and for this particular revisioning if Reich had used the case presentation—with its narrative point of view— of Ackerman and Strong? Similarly, could Brody have used Reich's radically subjective first-person account for a pluralistic ethics that requires the evaluation of conflicting moral appeals? Could Ackerman and Strong be as persuasive on behalf of a balancing approach using Brody's external focalization?

These questions hinge on the issue of the relation of the philosophical positions to the narrative positions. In order to examine these questions further, in the next chapter, I examine the kind of distance that exists between the various narrators of bioethics texts. This distance permits the author to add another form of commentary on the moral problems the case is supposed to simply illustrate.

Chapter Three: Distancing Oneself from the Case

"Because he did not consider himself to be the author of what he wrote, he did not feel responsible for it and therefore was not compelled to defend it in his heart."
—*Paul Auster*
CITY OF GLASS

Case 1: The Humane Murder of a Helpless Infant

Ms. R. arrived at the physician's office simultaneously depressed and agitated. She had been under Dr. T.'s care for a little over a year, but today was especially traumatic for her. Her breast cancer was developing rapidly. Since the birth of her daughter a few months ago, she had been rapidly losing strength. She feared that soon she would have to give up her baby in order that someone or some institution could care for her. She had no family, no husband, no one to turn to for help except Dr. T.

In the privacy of his office, she reviewed her tragic story. None of the obvious options would work. Adoption would not meet the baby's needs adequately, and given the baby's problem with a malformed hip, probably no one would want to adopt her anyway. As for the institutions that care for the moneyless orphans of the street, Ms. R. said

45

they could not possibly give her baby the care she needed. There were no friends or neighbors she could rely on. The city was large and anonymous.

She proposed to Dr. T. that she fulfill her responsibility to her child by doing what she considered the most living, the most caring thing. She wanted to bring about the death of the infant—humanely—so as to spare her the inevitable misery of her existence. She asked Dr. T. for information about how to carry out the fateful deed in a way that would be quick, sure, and peaceful. She was willing to take responsibility for her act. But given her own condition, time was short.

What the state might do to her, a dying woman, could be no worse than the death she faced soon in any case and the agony she would feel at failing in her responsibility to her child. She turned to Dr. T. for help.

THE NOTION OF THEORIES OF MEDICAL ETHICS

Before examining why key decision makers in this case reached the conclusion that the woman's plan was justified, it is necessary to have some framework for analyzing this problem and others of medical ethics.

The quotation above is from the first chapter of Robert Veatch's *A Theory of Medical Ethics*, titled "The Hippocratic Tradition." This book is considered by some as the first attempt in contemporary medical ethics to produce an extended theoretical framework for the discipline (Jonsen and Jameton 1627); in it Veatch expresses dissatisfaction with the Western Hippocratic tradition of medical ethics and wishes to construct a new moral framework based on social contracts. The quote above includes 1) the case presented at the beginning of the chapter (in italics as in the original), and 2) the first sentence of the next section, which analyzes theoretical· concepts. This quotation might at first seem an odd selection to make, but I wish to focus here on how Veatch has separated the case presentation from the analysis. This separation permits Veatch to frame the case and thereby signify to the reader that the text of the case should be treated differently from the subsequent philosophical analysis. Since Veatch does not provide a citation for this case narration, the reader assumes that he is the case's "author." In this chapter I wish to focus on this issue of authorship and its mirror reflection, the issue of readership.

Narrative theorists have been attentive to the issues of authorship and readership and have revealed that these issues are not as simple and natural as one might assume. At first, one might suppose that Veatch is obviously the author of this work. Narrative theorists wish to interrogate such an "obvious" reaction and query whether Veatch is the narrator of the book as well. Readers tend to be aware of the distance between first-person narrators and the author of a work, especially when there are obvious differences between the narrator and the author's life. Readers know that when Mark Twain uses the first-person in *Huckleberry Finn* there is a difference between the author of this work (Twain) and the narrator (Finn), a difference that Twain plays with in the beginning of the book. This difference can itself be an issue of debate, as seen in the controversy surrounding William Styron (an Anglo-American author) employing the first-person narration of an African-American slave in *The Confessions of Nat Turner*: Can an author create a narrator who is so historically and culturally distant from his contemporary world? Relatively self-conscious readers are aware of the difference between the author and the third-person narrator, especially readers who read various novels by the same author and become mindful of the different narrators used in each work. Readers are not generally aware, however, of how the concept of the author is itself a construct.

One can easily understand how different narrator personae can be assumed in various forms of literature. Most readers simply assume that the author of a work is the same as the actual person whose name appears on the work. Yet Mark Twain, for example, was the pen name used by Samuel Langhorne Clemens. Are these the same people? In order to clarify this problem, Wayne Booth in *The Rhetoric of Fiction* introduces the notion of the "implied author." As distinct from the "real" or the "biological author" (I am indebted to William J. Donnelly for this useful term), the implied author is the concept of the author that a reader receives from a piece of work. The biological author creates in the narrative "not simply an ideal, impersonal 'man in general' but an implied version of 'himself' that is different from the implied authors we meet in other men's works" (*Rhetoric* 70–71). So the image a reader attains of the author of *Frankenstein* is not the same as the biological author Mary Shelley, whose name appears as the work's creator. For Booth this implied author has distinctive moral consequences for our reading of a text, for it is the space between the implied author and the narrator that conditions our response to the moral import of a narrative. We encounter distinct "implied authors" in different texts that are tagged with the name of a particular biological author. This concept, Seymour Chatman notes, denies any "simplistic assumption that somehow the reader is in direct communication with (1) the real author ... or with (2) the fictional speaker, for how then could we separate the denotation (what the speaker says) from the connotation (what the text means), especially where these differ?" (*Coming* 76). The distinction between the biological and the implied author can be applied not only to fictional texts but also to philosophic ones. In the selection above from Veatch, we have two distinct texts, one separated by the other by different fonts; furthermore, we also have two narrators. We assume, however, that the two texts have a single implied author behind them.

We should also distinguish the biological author from what Booth later refers to as the "career-author." The concept of the career-author is created by readers in their encounter with the same "person" as they read different works which are ascribed to the same name (*Critical* 270–71). When we go to a bookstore and see a new book by the American writer Paul Auster, we bring with us assumptions of this work and its relationship to other works by Paul Auster. Auster has written a number of novels considered postmodern, most notably his subversion of classic detective fiction, *The New York Trilogy*. Yet he is also the biological author of a more conventional mystery novels, which he has written under a different name. The biological author is the same, but there are two career-authors. It would be possible for someone to read the differing career-authors and be unaware of their relation to each other. One could make a distinction between the ontological status of a career-author whose name is identified with a book's author (as "Paul Auster" is to biological Paul Auster) and that of the pseudonymous author of the detective fiction. The concept of various types of authors is useful in understanding academics such as Veatch, for there is a career-author behind his works who we come to assume exists. Consequently, in a bookstore we expect to find all the works on medical ethics by Veatch to be filed together. This is not simply a matter of listing items alphabetically, for if there were two ethicists named Robert Veatch, the bookstore owners would probably find a way to group the works by the same biological author together, that is, keep the two biological authors separated. We also assume that the philosophical ideas of the Veatch career-author will remain consistent and if there are any major philosophical changes then they would be noted as a dramatic change from earlier work. A work by Veatch that argued against contractual ethics and instead for a return to the Hippocratic tradition of paternalism would be seen as a remarkable alteration in the views of the career-author Veatch. Of course, Veatch could be associated with several career-authors. An author like Mircea Eliade, who wrote both works on comparative religion and novels, had two different—yet I suspect for readers often related—career-authors. The Veatch career-author is, however, more readily identified with the narrator of the philosophical passages than the narrator of the ethics cases. The identification we assume the implied author has with the philosophy-narrator rather than the case-narrator is important to explore further.

First, we should observe not only the differences in style between these two narrators but also the difference in their presence, for although there is a single philosophy-narrator in Veatch's text, there are a multitude of case-narrators. Each case seems to have an entirely different narrator. Compare the narrator of Case 1 to that at the beginning of Case 8:

> Researchers at a major metropolitan hospital and research center planned a cell physiology study requiring human endothelial cells. They proposed to obtain placentas from clinic patients in the delivery room for normal childbirth.
>
> The research proposal was submitted to the local institutional review board (IRB), the body responsible for review of research involving human subjects. The board first reviewed the protocol to determine, as required by federal regulation, if the risks to the subject were justified by the sum of the benefit to the subject and the importance of the knowledge to be gained; that is, the utilitarian test of maximizing net aggregate utility. They determined quickly that there was no plausible benefit to the subject of this basic science study, but they concluded that there was no plausible risk either. They were satisfied with the scientific merit of the research design and the importance of the research question and so determined that on balance the protocol passed the risk-benefit test. (190–91)

Clearly, this narrator seems more removed from the events and far less dramatic in recounting them than the narrator of Case 1. For example, in Case 1 the narrator makes unusual statements such as "The city was large and anonymous," while the narrator in Case 8 takes on the objective tone of a journalist simply recounting the facts: "They determined quickly that there was no plausible benefit to the subject of this basic science study, but they concluded that there was no plausible risk either." The freedom that ethicists have had in using different tones of voice for their narrators can perhaps best be illustrated by Veatch's own anthology, *Case Studies in Medical Ethics*, where almost every case presented has a distinct narrator.

Besides the style and presence, another difference between the philosophy-narrator and the various case-narrators is the degree of dramatization. Narrators who are dramatized make their presence explicitly known to the reader, yet the degree by which the reader is able to establish the narrator's attributes varies among texts (Booth, *Rhetoric*). Narrators may, for example, indicate their identity by revealing to readers their social status or political views, or narrators may be dramatized yet refuse to reveal their opinion on the events being narrated. In Veatch's book, the philosophy-narrator is dramatized throughout the work, a dramatization that leads the reader to believe that the philosophy-narrator and the implied author of the work are one and the same. In the second paragraph following the case presentation "The Humane Murder of a Helpless Infant," the reader comes upon the following sentence: "I shall use the term *medical ethics* to refer to a more general application of ethics to problems in the medical sphere, whether those problems are faced by lay people, public officials, physicians, or other health professionals" (16). In this use of the first person, we see the initial dramatization of the philosopher-narrator in Veatch's text. Yet the case-narrator is never dramatized, always left voiceless in the narrative presentation.

The various speakers in Veatch's text can be diagrammed in the following manner:

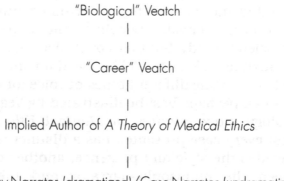

"Biological" Veatch

|

|

"Career" Veatch

|

|

Implied Author of *A Theory of Medical Ethics*

|

|

Philosophy-Narrator (dramatized)/Case-Narrator (undramatized)

These distinctions are important to the discipline of medical ethics, for the applied nature of the discipline makes the reader assume an intimate, "real" association between the biological author and the other levels of narration. As Booth observes, the associations among these various voices within a text are the subtext of a narrative presentation: "In any reading experience there is an implied dialogue among author, narrator, the other characters, and the reader" (*Rhetoric* 155). This dialogue is conditioned by the distance among the various voices. This distance can be physical, moral, intellectual, or aesthetic. Each of these distances in this dialogue influences how readers experience the text, and, for the ethics case, they affect how we judge the morality of the events.

Veatch's text also illustrates how in bioethics these relationships do not always form a simple series of distances between two participants in a dialogue, as suggested in Booth's analysis. Instead, the dialogue in bioethics can be a triadic one, for there are usually two different narrators: in the example from Veatch, the philosophy-narrator and the case-narrator. These various distances affect our experience of a case narrative. Since I suspect the reader tends to see the philosophy-narrator and the implied author as essentially agreeing with one another, the case-narrator exists as a distinct voice, separate and distant from the other two. An interesting exception to this general identification of the implied author and the philosophy-narrator can be found in Carl Elliott's article, "Philosopher Assisted Suicide and Euthanasia" where the philosophy-narrator proposes that those philosophers who support assistance by physicians in the death of patients be permitted themselves to carry out this action. Just like readers of Jonathan Swift's "A Modest Proposal," the reader discovers a moral distance between the implied author and the philosophy-narrator, and it is this distance that is the piece's true argument. The conventional identification of these two voices, however, and the customary separation of the implied author from the case-narrator reinforce the reader's initial experience of the case as objective and untainted by the positions of the philosophy-narrator. The various degrees of distance and identification that occur in bioethics have resulted in the ability of the biological author to weave a commentary into the events while seeming to remain an impartial observer. In order to demonstrate the importance of distance in the bioethics case narrative, I survey the way a moral subtext results in the various possible distances among authors, characters, narrators, and readers.

IMPLIED AUTHOR AND THE CHARACTERS

One of the most easily identified forms of distance is the one created between the narrator and the characters. Look, for example, at part of the narration concerning the Tuskegee syphilis study in the *Hastings Center Report*:

> Though the subjects had received the standard heavy metals therapy available in 1932, they were denied antibiotic therapy when it became clear in the 1940s that penicillin was a safe and effective treatment for the disease. Subjects were recruited with misleading promises of "special free treatment" (actually spinal taps done without anesthesia to study the neurological effects of syphilis), and were enrolled without their informed consent. Disclosure of the ongoing research in the popular media in 1972 led to termination of the study and ultimately to the National Research Act of 1974, which mandates institutional review board (IRB) approval of all federally funded proposed research with human subjects. (Crigger 29)

Here the distance between the narrator and the characters is obviously temporal, but it is also a distinctly moral distance as well. Interestingly the narrator uses the passive voice that hides the actors who are clearly portrayed (and I judge properly so) in a negative fashion. Here the judgments come in such choice of words as the subjects being "denied antibiotic therapy" which at that time had become "clear" was "safe and effective." The actors gave the subjects "misleading promises" and they are quoted as offering "'special free treatment'"; the narrator then notes in parenthesis that this free treatment was not a "treatment" nor was it without discomfort. The narrator clearly is revealed as having a moral distance from the actors in this narrative. It is a distance, I suspect, shared by the implied author and the reader. This distance becomes confirmed by noting how the public report led to the termination of the study and the creation of boards of review to prevent these events from occurring again.

IMPLIED AUTHOR AND THE UNRELIABLE NARRATOR

> Mr. X, a fifty-four-year-old patient with a long history of nodular goiter, presented with recent growth of thyroid mass and hoarseness. Surgery was performed following a biopsy diagnosis of anaplastic thyroid carcinoma. Only partial removal of the tumor was possible, however, and pulmonary metastases were also suspected. Three weeks after the operation, the patient complained of dyspnea, which proved to be caused by a local recurrence. Further tests revealed lung and bone metastases. Despite irradiation and chemotherapy, as well as palliative treatment, the patient died about three months after the initial diagnosis.

This is the beginning of Bettina Schoene-Seifert and James Childress's (or perhaps one should say the philosophy-narrator's) consideration of the question, "How much should the cancer patient know?" The reader does not encounter an overt ethical dilemma until the second paragraph of the case narration.

> The patient, a hard-driving entrepreneur who dominated both his family and his business, was first told that there was a probability of "malignant cell transformation" in his thyroid gland. He immediately consented to the recommended surgery and was told afterward that the diagnosis had been confirmed and that the tumor had been successfully removed. The likelihood of lung metastases was not mentioned.

After mentioning that Mr. X was not told of his condition, the case-narrator relates that, "During a two-hour conversation, the physician then informed the patient's wife, son, and daughter-in-law of the patient's clinical status and his extremely bleak prognosis." The irony of this presentation lies in the case-narrator's description of the physician's conversation with the family in the manner that the reader would expect of a physician informing a patient. "The physician encouraged them to ask everything they wanted to know and answered every question with patience and empathy." Because of our foreknowledge of Mr. X's death, the reader witnesses as all the patience and empathy that we would expect to be given to Mr. X, is given to his family. The case-narrator goes on to describe the family's decision not to inform Mr. X of his condition, and the physician's acceptance of this decision. Booth has described this type of narrator as "unreliable," that is, the narrator speaks or acts contrary to the moral norms of a work (*Rhetoric*). This incongruity of norms is one way for the biological writer to provide a commentary upon the moral nature of events. The reader comes to conclude that this case-narrator is unreliable, for this "informed consent" is a parody of the traditional moral values embedded in informed consent. The case-narrator's unreliability is continued when the reader then learns that "Mr. X was told only that he needed 'preventive' treatment. After being informed of possible side effects, he enthusiastically consented to irradiation and chemotherapy." Here the use of the phrases "being informed" and "enthusiastically consented" makes the reader see the case-narrator as having differing norms from the implied author. The case-narrator then describes the treatment of Mr. X in contrast to the family. "The family inquired daily about the patient's condition, either by phone or in person, and was treated very kindly by the staff.... In the hospital, the patient's complaints and harsh criticism of the nurses generated hostility, causing them to avoid him whenever possible."

When the philosophy-narrator comments upon the scene presented, we see how this narrator's norms agree with those of the implied author. The philosophy-narrator makes statements that are not so much arguments as presentations of previously acknowledged principles to a fellow insider. Neither the case-narrator nor the characters in the story are able to see either the irony of the situation or the principles that make their actions a parody of the norms of medical ethics. The reader is expected to have been outraged—as the philosophy-narrator is—by the events narrated: "It is a form of insult and disrespect to abridge a competent patient's rights when the only reason is to protect what others regard as the patient's own welfare." The unreliability of the case-narrator leads the reader to agree with the philosophy-narrator's gloss on the events, for the implied author controls the entire text and prepares the reader to find the case-narrator unreliable and, thereby, the moral position of the philosophy-narrator reliable. At the heart of this difference is the seemingly unintended irony of the case-narrator's presentation of the informed consent. The selection of a narrator who seems oblivious to the irony of describing the events in the language of informed consent permits a commentary on the events by these biological authors.

IMPLIED AUTHOR AND THE READER

In 1988, the series "A Piece of My Mind" in the *Journal of the American Medical Association* printed an anonymous article titled "It's Over, Debbie." This essay has become a key case for medical ethics, often cited as an instance of the dangers in the practice of active euthanasia. Albert Jonsen has used it as an example of a case that can be resolved with casuistic ease ("Casuistry"). Howard Brody analyzes it through the tension between care and work in health care (*Healer's*). There has been much debate about the authenticity of this narrative and subsequently the editors' intentions in printing it. Some have claimed that the narrative is a pure fiction and thereby not worth our attention except perhaps as a speech act in the medical ethics community. Although this does raise once again the importance of the "real" within medical ethics, for our concern here it raises questions about the status of the reliability of the implied author and in the end, considering the criticism of the case, it reveals the distance between the norms of the implied author and those of the readers.

The article, which is merely three paragraphs long, is told by a dramatized narrator:

> The call came in the middle of the night. As a gynecology resident rotating through a large, private hospital, I had come to detest telephone calls, because invariably I would be up for several hours and would not feel good the next day. However, duty called, so I answered the phone. A nurse informed me that a patient was having difficulty getting rest, could I please see her. She was on 3 North. That was the gynecologic-oncology unit, not my usual duty station. As I trudged along, bumping sleepily against walls and corners and not believing I was up again, I tried to imagine what I might find at the end of my walk. Maybe an elderly woman with an anxiety reaction, or perhaps something particularly horrible.

One can hypothesize (as Kathryn Montgomery Hunter has) that there is a large degree of temporal distance between the implied author and the narrator, but the moral and intellectual distance seems close. After meeting with the dying woman, "Debbie," the resident provides a graphic portrayal of the scene:

> The room seemed filled with the patient's desperate effort to survive. Her eyes were hollow, and she had suprasternal and intercostal retractions with her rapid inspirations. She had not eaten or slept in two days. She had not responded to chemotherapy and was being given supportive care only. It was a gallows scene, a cruel mockery of her youth and unfulfilled potential. Her only words to me were, "Let's get this over with."

Here with the observations of the "cruel mockery" and the "unfulfilled potential" the implied author and the narrator seem close to one another. If the implied author does not completely approve of the views and actions of the narrator, the design is to make the narrator reliable. That the judgment by most readers of this case was that the resident acted immorally, even if they supported active euthanasia, provides an example of the kind of moral distance that can occur between reader and implied author, for it is the construction of this reliable narrator that makes this distance possible. If the biological author had constructed this narrative so that there was some form of moral distance between the narrator's perspective and the implied author's then readers would not have reacted to the case in the way that they did (cf. Abbott).

ENTRAPPING THE CASE'S NARRATEE

Those already aware of the various levels of authors and narrators within a piece of fiction might perhaps be critical of my naive and uninflected use of the term "reader" throughout this analysis. Narrative theorists have been aware that not only are there different levels of senders in a story but also different levels of receivers. To simply refer to the receiver as a "reader" joins together elements that should be separated out, for each of these constructed senders are addressing different constructed receivers. I am not merely speaking of the historical and social distance between, for instance, a twentieth-century American reader and a nineteenth-century French novelist, but of less obvious differences among the imagined receivers of the messages of the biological author, the implied author, and the narrator. At times when we read a narrative, we have a distinct sense of who these various receivers are suppose to be. Gerald Prince has examined in detail the various types of narratees in order to demonstrate the complexity that exists in this concept ("Introduction"). Just as biological authors create implied authors and narrators for their works, they also construct implied readers (see Iser; Chatman, *Story* 149–151) and narratees. We can revise our diagram of the various senders of a message in bioethics to include various receivers of the message (cf. Martin 154):

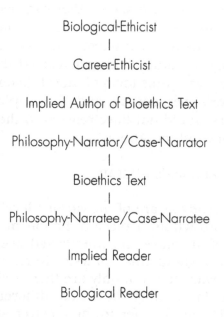

Biological-Ethicist

|

Career-Ethicist

|

Implied Author of Bioethics Text

|

Philosophy-Narrator/Case-Narrator

|

Bioethics Text

|

Philosophy-Narratee/Case-Narratee

|

Implied Reader

|

Biological Reader

In the issue of the ethics case narratives, this means that we should attend to the various receivers, including the difference between the imagined receivers for the case versus the imagined receiver for the philosophy. Look at the following example, which has a dramatized narratee:

> You are a psychiatrist for the student health service at a large university. One of your patients is a foreign graduate student who has been extremely depressed since his affections were rejected by his former girlfriend, who lives in the city with her parents. Today, expressing his anger to you over the way the girl treated him, he announces his intention of killing her when she returns from a trip out of the country.
> You know that the business of predicting violent behavior

is not as clear-cut as some would like to think; nevertheless you feel that your patient is seriously disturbed and that there is a fair probability that his threat will be carried out.

Many will recognize this case as a retelling of "Tarasoff v. Regents of the University of California." It appears in Howard Brody's bioethics textbook *Ethical Decisions in Medicine*. This case concerned Prosenjit Poddar, who killed Tatiana Tarasoff after revealing his intentions to a psychologist at the University of California at Berkeley. Bioethicists often cite this case as a classic demonstration of the conflict between patient confidentiality and the physician's obligation to protect society. As in the majority of his case presentations throughout this introductory bioethics textbook, Brody employs the second-person narrative voice and concludes the presentation with a question demanding response from his reader: In this case he asks, "What should you do with your knowledge?"

Compare Brody's presentation with that of Beauchamp and Childress in *Principles of Biomedical Ethics*:

> On October 27, 1969, Prosenjit Poddar killed Tatiana Tarasoff. Plaintiffs, Tatiana's parents, allege that two months earlier Poddar confided his intention to kill Tatiana to Dr. Lawrence Moore, a psychologist employed by the Cowell Memorial Hospital at the University of California at Berkeley. (3rd ed. 400)

Beauchamp and Childress's version maintains the classic third-person "just the facts" orientation of many principle-centered bioethics discussions. The authors note that they have maintained the "court-style" presentation, and, by doing so, they place the reader in the position of the judge. This judicial "objectivity" is characteristic of principle-centered narrative presentations. Brody, though, does not begin with the dramatic events of Tarasoff's death, but instead first creates a distinct narratee to whom his narrator speaks. While in

Beauchamp and Childress the narratee has a role within the narrative event—a part in the ethics discussion—Brody positions the narratee within the flow of the past events, that is, in the narrated event. In Brody's there is no place for the narratee to escape or hide. The sense of being trapped and placed within the narrative's dramatic pulse is the result of this use of the second-person voice. As Gary Saul Morson writes, "The second-person address, constantly threatens to make the reader a character in the story, a participant in an unedited dialogue with the author—a threat that is the first of the work's many strategies to overcome the aesthetic distance it assumes we 'assume'" (382). Reading Brody's presentation we are not with the narrator (as we are in Beauchamp and Childress). Instead, we are the psychotherapist.

While the second-person voice is common in bioethics case narratives, its use is unusual in most literature, for it has an intimacy difficult to maintain in an extended prose piece. Leo Tolstoy uses it in his short story "Sevastopol in December," and the Latin American writer Carlos Fuentes has employed the second person in the novella *Aura*. In these instances the writers entrap the "reader" in the narrative, placing the reader in the unusual position of being a part of the plot. In Fuentes's *Aura*, for example, a young man (with us, the readers) answers an advertisement for a job and finds himself (with ourselves) reliving a cyclical nightmare. Fuentes is able to play with the identity of the main character in the story by implicating the narratee as simply another person in this horrific cycle, and he draws the narratee into the narrative as he directs not only what the narratee sees and does but also how the narratee feels and thinks about the situation. Like Fuentes, in the Tarasoff case Brody is able, through the second person, to ascribe thoughts and feelings to the narratee, to entrap the narratee in the situation. For example, in Brody's statement, "You know that the business of predicting violent behavior is not as clear-cut as some would like to think," he seductively leads the narratee to agree to his claim about the possibility of this man doing violence to the woman, for the narratee is shown to be more sophisticated in his or her thinking than "some." Unlike Beauchamp and Childress, Brody takes the narratee inside the therapist's mind, and even more disconcerting is that in Brody's account the therapist's mind is the reader's mind.

The two Tarasoff case narratives also differ in the way events are ordered. Beauchamp and Childress begin with the result of another's actions, while Brody wishes the narratee to make a decision about an action *to be* taken, that is, the point at which the therapist had to make a decision without the luxury of knowing the results of that decision. Later Brody provides us with the conclusion of the Tarasoff case, but his presentation recreates the dilemma in terms of how physicians come across ethical dilemmas, that is, as personal moments of confrontation. Beauchamp and Childress's presentation, by contrast, is the way in which philosophers encounter ethical issues, as past events that can be judged and do not require the philosopher's action.

Brody is explicit at the beginning of his textbook that his objective is not to offer "a set of rules," "right" answers, "a code of ethics" or "medical law" (xvii). Instead he wishes to provide a rational method for bioethical decisions. Unique to Brody's discussion of bioethics is the emphasis on "action" and understanding the consequences of those actions. Brody notes that after we are able to recognize the problem, "The second major ingredient of an ethical problem is that the person involved must place a significantly different value upon each possible action or upon the consequence of that action" (5). This feature is emphasized in Brody's style of presentation; his case narration forces us to make decisions about what we should do and not to reflect on the principles upon which we judge *other's* actions. Brody places the reader within the case at the point where he wants students to make ethical decisions, not as outside, third-person judges but as involved actors, who must predict and take responsibility for the consequence of their actions.

This chapter has been an exploration of the complex interaction of various senders and receivers—imagined and real—in a bioethics text; each distance created by the author accordingly permits an insertion of a moral judgment on the events presented. The temporal distance between the narrator and the characters in the Tuskegee syphilis study description translates into a moral distance as well, for the narrator is able to identify the participants' actions and decisions with a critical eye. The construction of an unreliable narrator in the case of Mr. X allows the author to create a moral distance from the values of the implied author. The distance experienced between the implied author and the reader of the Debbie case results in the strong negative—and probably unintended—reactions by many of the case's biological readers. Finally, the closeness of the narratee and the character within Brody's version of the Tarassoff case permits the author to entrap the implied reader of the case within the narrative, and, thus, this narrative strategy forces the reader to respond personally to the moral quandary presented. These distances are part of all narratives, and Booth was particularly attentive to the moral dimensions within the adoption of particular types of implied authors in novels. The presence of these various distances within bioethics is particularly important because of the explicit moral objectives of presenting these narratives. Readers of bioethics need to be attentive to how "commentary" can subtly be inserted within the description of events.

Chapter Four: The Chronotope of the Case

Does the category of "setting" even have any particular meaning for the bioethics case? Mieke Bal points out that there is always a location for events in a narrative, and if one is not supplied by the narrator, the reader will create it for the story. She suggests that one can mark the location in each story and then "investigate whether a connection exists between the kind of events, the identity of the actors, and the location" (43). Genres have long been defined in terms of their sense of place. For instance, contrast two classic genres: the Western's single-street towns (which are dwarfed by the open landscape) and the future dystopia's decayed city (from which one cannot escape). The sense of space conveyed in these genres in turn determines the kind of action that can take place within them. The genre of the ethics case seems to be one that has more in common with those genres tied to the urban environment than those set in the rural world. In the idealized world of the country general practitioner, there was no need for academic philosophers to assist with issues of confidentiality, while the urban physician needs the help of other specialists in a world dominated by, and in some manner made through, specialists.

Consider, for example, Robert Veatch's case "The Humane Murder of a Helpless Infant," already discussed in relation to the question of authorship. The urban world has a high degree of salience in Veatch's narrative. Look at the following selection of his narration:

> In the privacy of his office, she reviewed her tragic story. None of the obvious options would work. Adoption would not meet the baby's needs adequately, and given the baby's problem with a malformed hip, probably no one would want to adopt her anyway. As for the institutions that care for the moneyless orphans of the street, Ms. R. said they could not possibly give her baby the care she needed. There were no friends or neighbors she could rely on. The city was large and anonymous.

Veatch opens this case with Ms. R.'s arrival at Dr. T.'s office, and he quickly juxtaposes the intimacy of Dr. T.'s office with the hostile city. Outside of the safe space of this office lies an urban world with its "institutions," "moneyless orphans of the street," and "the state." As Veatch pronounces in a matter-of-fact tone after the second paragraph, "The city was large and anonymous." It is not her impending death or the nature of her baby's needs that are the source of Ms. R.'s dilemma but the dangers of this uncaring urban world. She desperately turns to the only man she can trust, Dr. T. Through this dramatic presentation Veatch sets a distinctive stage for this dilemma. He is not merely conveying the context of the problem, but fabricating a world in which his characters act. Though we may accept that this view of the city as a dangerous place may be a reflection of Ms. R.'s perspective, the omniscient narrative voice's pronunciation of the city as "large and anonymous" affirms this judgment.

Veatch's hostile urban world has similarities to the literary landscapes of the hard-boiled detective novels of Raymond Chandler and Dashiell Hammett. In this

American literary genre, the heroes are private investigators, outsiders to a cold, corrupt world. These hard-boiled detectives act in an immoral (or at least amoral) urban domain. The detectives have private moral codes that run against those of the corrupt world they must live in. Anti-heroes such as Sam Spade, Mike Hammer, and Philip Marlow begin their adventures reluctantly; they are suddenly confronted with a "case." These cases concern individuals who lack trust in the mechanism of the urban social world. These desperate men and women stumble into the detective's office, having nowhere else to turn. Similarly, Veatch opens with Ms. R.'s dramatic arrival in Dr. T.'s office. She has "no family, no husband, no one to turn to for help except Dr. T." The urban institutions—like the police in the hard-boiled narrative—cannot give her assistance; Ms. R. states that she has no friends that she can trust with her child's care. As in the world of the hard-boiled detective, Veatch's character must go outside the law for help. The woman is trapped in a city where she lacks communal and familiar connections. Veatch, who in this work argues for the use of social contracts, here depicts a world where we cannot rely on established social agreements concerning moral actions; we must, in order for justice to be carried out, make confidential pacts with *private* investigators. As Chandler noted, the emotional grounding of this style was believing that right will triumph over wrong only if "some very determined individual makes it his business to see that justice is done" (2).

The degree of particularity of place is also something worth noting. Although there is a distinct sense of place in Veatch's case, is there a sense of a particular urban space or does it represent simply "urbanness"? Mikhail Bakhtin has observed that in the classical romances, events could be taking place in almost any city of that period (100), and in Veatch, too, the lack of the particularity of place is noteworthy. Contrast this with how many contemporary novels such as *The Moviegoer* and *Midaq Alley* depend on a sense of place (Percy's New Orleans and Mahfouz's Cairo) for their narratives to cohere. Veatch's case is not unique; in general, most bioethics cases simultaneously have a distinct sense of place but do not designate a particular location. Events take place in the contemporary clinical environment yet not any particular clinic. While some might contend that these generic medical settings must be designated as such in order to preserve the confidentiality, this lack of distinct place has a profound influence on how the reader experiences the narrative, one that is similar to the manner Dena Davis has mentioned concerning altering biographical details. Cases such as the Dax Cowart case (which I will examine in more detail in following chapters) have elements that make them distinct as much for their place as for their characters. Cowart was severely burned in an accident and requested that care for his burns be stopped. Although little is mentioned of the environment where these events occurred, taking into account the ethos of East Texas, which has distinct concepts of gender, individuality, and religiosity, makes the shape of the narrative more plausible than if the events had taken place in New England.

Another feature of this lack of particularity is that the problems of social justice simply lose their meaning. Even within a single city, understanding setting can be essential for raising questions concerning economic justice. For instance, in Chicago, a hospital that must respond to the needs of a homeless person will in turn deal with this issue differently if it is in the wealthy northside or in the poor southside. Without a sense of place, the reader cannot know the limitations on action and the expectations for particular types of actions.

The Veatch case does provide a sense of location and

of movement between locations. We begin with Ms. R.'s arrival in Dr. T.'s office. Bal observes that stories often exist in spatial oppositions, such as inside/outside, center/periphery, and high/low. With each of these spatial oppositions come oppositions of ideas. As in many ethics cases, the location of events in the Veatch case moves between two places: the non-specific "home" and the clinical setting. These binary oppositions also mark the two places in terms of private versus public. Stories of sickness become ethics cases when they move from the home setting to the clinical setting, for in doing so the events move from the private sphere to the public one.

Interestingly, bioethics as a discipline is characterized by this change in setting. There are many causes for the rise of bioethics as a distinct discipline (and distinct from the Hippocratic tradition), but one notable element has been the problems that arise through a similar change in the setting for the practice of medicine. The Hippocratic tradition represents the environment of the home. As the Oath states, "Whatever houses I may visit, I'll come for the benefit of the sick, remaining free of the intentional injustice, of all mischief and in particular of sexual relations with both female and male persons, be they free or slaves." This is an image of an outsider invited into the private realm. The contemporary movement in medical ethics, referred to as "bioethics," represents the move of the patient to the public domain of the clinic. David Rothman characterizes the changes of medical decision-making in the United States as the coming of "strangers at the bedside" and this was first seen in "a new commitment to collective, as against individual, decision-making" (3). This collective decision-making, which includes lawyers and philosophers, is part of a change in the public world of medical care, a radical transformation of setting.

TEMPO AND THE TIME WARPING OF THE CLINICAL SCENE

Literary genres can be distinguished not only by their various uses of space but also by the way time is manipulated. For example, a novel is distinguished from a short story not in terms of how language is used but in terms of reading time. One of the expectations of a bioethics case is that it will take a "short" period of time to read or hear, far shorter than a short story, perhaps the length of a parable. With attention to this feature of a narrative, one can, however, compare the relationship of the time narrated to the time of narration. The difference between the two profoundly affects our experience of different narratives. Think for example how *Citizen Kane* plays with various forms of time. The initial scene of Kane's death is in slow motion, so that it takes longer to watch the events then they occur in the narrative time. In a later series of scenes, we watch as Kane's relationship with his wife passes from close affection to cold distance over a period of probably decades yet the time for the viewer is only a matter of minutes. Finally there are entire scenes of the film that show us time in such a manner that we assume that we are experiencing it at the same time as the characters.

It is harder with written narratives to compare reading time with narrative time. In the conventions of reading comics, there is an ongoing and immediate relation of space to time, as the physical frames of the comic can be related to the movement of time in the story (McCloud 100–102). How does one classify different rates of reading a literary text (cf. Genette *Narrative*; Rimmon-Kenan)? One can compare, however, the relation "between duration in the story (measured in minutes, hours, days, months, years) and the length of text devoted to it (in lines and pages), i.e., a temporal/spatial relationship" (Rimmon-Kenan 52). Look for example at the following case:

Michael Marks, age 55, was admitted to the hospital because of shortness of breath. For 15 years he had suffered from chronic bronchitis with emphysema, complicated by intermittent brief lung infections. He also suffered from chronic pain in his shoulder muscles. This pain had been thoroughly investigated on a previous hospital admission; no cause was found.

Mr. Marks was admitted to the intensive care unit because of his precarious respiratory function. He did not require the use of a respirator. He repeatedly and loudly called out to the nurses, disturbing other patients, asking for codeine injections by name. His physician was reluctant to give him narcotics because of his respiratory status. Narcotics suppress the brain's respiratory drive center. Instead he ordered injections of salt water.

Mr. Marks received several injections at three-hourly intervals. He no longer complained of pain and slept apparently well. (Abrams and Buckner 619)

In this case presentation, an entire night passes but the narrative is told in [eighteen] lines. (All my references to textual space apply to the text above and not to the original version.) Case presentation can be longer, although to do so is to break from the genre's conventions (e.g., Miles and Hunter). But I wish here to argue that one of the defining features of the bioethics case is not the overall period of reading time but a particular textual tempo. Through a study of the bioethics case's tempo, one discovers how the narratives focus primarily on events within the space of the clinical setting.

The tempo of a narrative is generally one of either acceleration or deceleration (Rimmon-Kenan 52–53); time possesses extraordinary plasticity within the text. In the second, third, and fourth lines of the case concerning Mr. Marks, fifteen years of his life are summarized: "For 15 years he had suffered from chronic bronchitis with emphysema, complicated by intermittent brief lung infections." In a similar manner, the time of the story as compared to the space of narration in the final lines takes place over an entire night: "Mr. Marks received several injections at three-hourly intervals. He no longer complained of pain and slept apparently well."

A more detailed look at how these variations in tempo occur will be helpful in understanding the tempo of the ethics case. Acceleration or deceleration can be accomplished in five basic ways: ellipsis, summary, scene, stretch, and pause (Genette *Narrative*; Rimmon-Kenan).

Ellipsis in the tempo occurs when events are entirely omitted from the narrative space. In the Marks case, we learn that he received injections but the time between the injections is completely passed over in the narrative.

Summary, the second type of tempo, consists of the narrated events being contracted into a small amount of narrative space. The first four lines of the case represent a summary in that Mr. Marks's admission is condensed into that first line and then fifteen years are compressed into the second, third, and fourth lines.

The third type of tempo, scene, represents events in a manner close to the time of the story. The most common form of scene presentation is in a piece of dialogue, where the time it takes to read the dialogue is approximately the same time it would have taken to listen to the characters speaking. A good instance of scene is the film *High Noon* where the film time and the narrative time are exactly the same, so one could start watching the film at the same time as the time in the film and the clocks of the town and the clocks of the viewer would continue to be in sync. There are no moments of scene in the Marks case.

The fourth type of tempo is stretch, in which the description takes a longer period of time than the events of the story. In this tempo, we move away from the one-to-one ratio of tempo in scene and toward the narrative event taking longer than the narrated event. Perhaps the most well-known example of stretch is Joyce's *Ulysses*, in which the action of a single day can take months for a reader to experience. Repeated representation of events can also create a form of stretch. Quentin Tarantino's *Jackie Brown* repeats a single event through the perspective of different characters. The narrative event becomes slowed and repeated in a manner that causes it to take three times longer than the narrated event. I am not aware of an ethics case at present in which the stretch tempo is employed.

Pause is the final form of tempo used and it consists of story space in which no time passes in the story. In this form of tempo the narrated events stop but the narrative event continues on. In film this can be dramatically used as in the instance of a film like Francois Truffaut's *Jules and Jim* in which the director uses sudden stills to stop time in the narrative. In literary discourse the use of description of a scene can also stop the time of the narrated events. In the case above, the line "Narcotics suppress the brain's respiratory drive center" occurs in the middle of the description of events and in doing so causes the narrated events to pause to explain this additional material. If this had been an instance of reported speech instead of a presentation by the narrator, the tempo of the story would be an example of scene.

The tempo of bioethics cases tends to move between summary and ellipsis. As in the medical case history, there can be space devoted to descriptive pauses, primarily because it is the body that is being described. Some case presentations try through dialogue to create scene. John Lantos in "Leah's Case," a case concerning the medical treatment of a Jewish Orthodox eighteen-year-old, writes the case primarily in dialogue that draws most heavily on scene. The ethics committee discusses whether the woman should be told of the need to remove a tumor, news which may result in her refusal to have it done.

> "Her life might not be worth living if she is sterile. That should be her decision to make. If she chooses to die rather than be infertile and live, her choice should be respected. But she can't make that choice unless she knows what we all now know. She must be told and must be allowed to decide."
> "How can an 18-year-old make a decision like that?"
> "According to the law, she can make it."
> "According to your law. Not according to her law."
> "She would not have to consent to treatment. Parental consent would be sufficient."
> "What if she was 17 instead of 18, and was here?"
> "Legally, her parents could decide." (83)

Although the tempo of Lantos's case presentation consists predominately of scene, it also has periods of ellipse and summary. And the points in the story at which the tempo enters into ellipsis rather than summary and scene are a telling feature of bioethics case presentations.

THE CHRONOTOPE OF THE CASE

Bakhtin has noted that genres can be distinguished by their differing use of what he termed "chronotopes." Chronotopes are the interrelations of time and space in narrative discourse. Drawing upon Einsteinian physics, Bakhtin claimed that time and space must be understood as a single unit and thereby did not wish to separate the analysis of time from space in the analysis of literary genres. "In the literary artistic chronotope, spatial and temporal indicators are fused into one carefully thought-out, concrete whole. Time, as it were, thickens, takes on flesh, becomes artistically visible; likewise, space becomes charged and responsive to the movements of time, plot and history" (85). The relationship between the chronotope and action is not simply that the actions occur within a chronotope but rather that the chronotope conditions the type of actions that can take place (Morson and Emerson 369). Chronotopes, in Bakhtin's view, are not just an additional component to understand plot, for they shape plot by shaping time and space. Thus the chronotope of the cyberpunk world as seen in Ridley Scott's film *Blade Runner* and William Gibson's novel *Neuromancer* is the result of combining the dystopian future time of science fiction and the cityscape space of the hard-boiled detective novel. The result is that the types of actions that occur focus on morality and identity.

Distinct arenas of life can be occasions for chronotopic motifs, which in turn can necessitate particular events. Bakhtin gives the example of the chrontope of the road, where "the spatial and temporal paths of the most varied people—representatives of all social classes, estates, religions, nationalities, ages—intersect at one spatial and temporal point" (243). The chronotope of the road in turn necessitates stories of encounter because of the dissolution of the normal social distances in various social groupings. Time in this chronotope becomes influenced by the spatial elements of the road and thus often involves luck and chance. Another example Bakhtin gives is the castle chronotope of the Gothic genre. He sees the physical space of the castle as being charged with the historical past. Everything in the castle is a signifier for another time, as they are filled with the actions and events of past family members. A castle that was supplied with contemporary furnishings would lose the chronotope of this genre and thus would lose the plot as well.

I mention the chronotopes of the road and castle, for the genre of the bioethics case also possesses its own unique chronotope that depends upon a unique chronotopic motif—the clinic. In the genre of the bioethics case, the tempo of the discourse is often directly related to the patient's entrance into the space of the medical world. The farther the character goes from the medical world, the greater the chance for ellipses. Of course, the tempo of the discourse is always related to the reportability of events, for there can be extensive medical care condensed in a single sentence, but the tempo rarely includes periods when the characters outside the space of the medical world discuss problems and thus provide a greater degree of space in the discourse. In other words, the tempo of the narrative expands as the patient enters into the sphere of the health care professional. The physical space of the text increases as the characters enter the space of the hospital or medical office and decreases when the events of the story take place outside this setting.

The consequence of this chronotope is that social issues tend not to be a concern of medical ethics. When cases are presented with a chronotope that extends our moral concerns only toward what is happening within the setting of the hospital, the result is that the world outside the hospital does not seem to matter. Characters' actions have significance only within the setting of the clinical environment. The chronotope of the clinic in bioethics cases is also unique in its lack of historical time. This is a space that, unlike the Gothic castle, seems to exist without having a time outside of the pathographic time of the patient's life. The clinic does not reside within larger narratives of community life or intersect with the socioeconomic life of the city. Instead, the chronotope of the clinic seems to exist in another time and space apart from the world of the larger environment of the city or neighborhood.

In the chronotope of the hospital, people are usually transformed into particular roles. One must, as in a preset drama, take on one of a set of predetermined roles once one has entered the clinic: physician, nurse, patient, technician, or visitor. Character roles are thus defined once people enter the chronotope of the clinic.

I began this chapter with Bal's observation that all narratives need to take place "somewhere": We cannot conceive of stories as simply happening outside of a particular setting, whether it is furnished by the narrator or is created by the reader. This chapter has been an exercise in taking the scene in which events occur within the case as a serious category for understanding the rhetoric of bioethics. Veatch's creation of a particular hostile urban world is itself part of his argument for a contract-based ethics for medicine. One need not make new moral contracts for society if we lived in a environment where we know everyone and are known by everyone. Furthermore, the ethics case tends to create a relationship between the space devoted in the physical text to the space in which events occur in the story. In a type of literary relativity, the space of the text expands (and with it the time for reading) as the story focuses on the clinical setting and the text contracts when events take place outside of that setting. The chronotope of the bioethics case (its unique merger of time and space) is the world of the medical clinic. Although it may seem understandable that bioethics cases take place within the clinical setting, this "natural" setting has two consequences for how medical ethics is conceived. First, our attention becomes focused, literally and philosophically, on how problems are framed in the clinical setting, and we are not asked to imagine how the events outside of this setting influence the shaping of moral problems. Second, the tendency for the creation of "generic" clinical environments within ethics cases has resulted in a lack of concern for the particularity of place in bioethics. The lack of particularity has, I suspect, led in turn to the field's general disinterest in issues of social justice. What seems necessary for bioethics is that to expand the philosophical standpoint one needs to expand and to focus one's physical standpoint.

Chapter Five: Opening and Closing the Case

> "'Begin at the beginning,' the King said, gravely, 'and go on till you come to the end; then stop.'"
>
> *—Lewis Carroll*

In *The Sense of an Ending* Frank Kermode argues that we use narrative as a way of humanizing time, a way to battle against the innately unstructured manner in which the world has presented itself to us. Kermode offers the ticking of a clock as an example of this attempt to create order out of the chaos. He notes that when asked what a clock "says," we often reply "tick-tock," and "it is we who provide the fictional difference between the two sounds; *tick* is our word for a physical beginning, *tock* our word for an end" (44–45). For Kermode this tick-tock is a model of plot, "an organization that humanizes time by giving it form" (45). It is not as simple as the King suggests in Lewis Carroll's tale. Isn't the very problem knowing what is the beginning and what is the end? The world does not manifest itself to us as a series of tidy stories, for we must humanize our world by plotting it. In this chapter I examine the tick and tock of the bioethics case. The genre of the ethics case is partly defined by its distinct definitions of opening and closing, and these have implications for what we think the data of bioethics is and what we should do with this data.

OPENING THE CASE

One of the principle features that distinguishes a narrative from a nonnarrative is temporal disruption; in a narrative something "happens." Jurij Lotman, in *The Structure of the Artistic Text,* argues that we can divide texts into "those with plot and those without plot" (236–39). For Lotman, texts that lack a plot bear a "classificatory character." One of such a text's properties is that it consists of a particular way of ordering the world, and "it does not permit its elements to move in such a way as to violate the established order" (237). Paul Ricoeur remarks, concerning Lotman's text without plot, that it is "a purely classificatory system, a simple inventory—for example, a list of places, as on a map" (167). A text *with* plot entails a disruption of a classificatory system: "The movement of the plot, the *event*, is the crossing of that forbidden border which the plotless structure establishes" (238). In other words, an event occurs and thus becomes what narrative theorists refer to as "reportable" when some form of transgression takes place (Prince, *Dictionary*).

Readers have different expectations of reportability with a particular genre, for each genre represents a disruption of different classificatory fields, and a reader when given a particular genre anticipates a distinct type of transgression. For instance, suppose you go to a bookstore and find an interesting book in the section marked "murder mysteries," which you buy and over the next week read. This novel entails family conflicts, some turning quite violent, but no one is killed.

Although you found the story interesting, you are disappointed, but what is the reason for your disappointment? Because the story does not have the reportability of the murder mystery genre, it has not fulfilled your expectations. Suppose that in this novel the violence does result in the murder of the father by his brother, and this is witnessed by the entire family. In such an instance, you are still disappointed, because you read about a murder but there was no mystery. Reportability is therefore not self-evident but must be placed within the context of the genre, that is, the reader's expectations of a particular form of transgression (Lotman 234).

Ethics cases and medical cases differ in terms of their reportability. An interesting example of the differences in reportability can be found in a case presented in a *New England Journal of Medicine* article, "Basic Curricular Goals in Medical Ethics." Culver et al. argue that "medical students should be able to identify the salient moral components of" the following "case":

> A competent patient with an obvious malady consults a physician who suggests a treatment that will almost certainly be effective. The physician informs the patient that the treatment has one minor risk; the patient asks a pertinent question about that risk and then decides to proceed. The treatment is carried out, and the patient recovers from the malady. (34)

The case has reportability (and "salient moral components") but it is not the reportability of an ethics case. Instead it is the reportability of a medical case, but I imagine for most clinicians a very dull one. The reportability of the medical case is the result of a transgression in a person's "previous state of health"; the reportability of the ethics case arises from a transgression in morality. The reader has these differing expectations of reportability when they come upon cases defined either as "medicine" or "ethics." Ethicists have their own criteria of an "interesting" case, which is qualitatively different from an interesting medical one. A single case may be interesting for both clinician and ethicist, but this would be for different reasons. The problem with the "Basic Curricular Goals" case is that it lacks a moral violation and thus—like the murder mystery without a murder or a mystery—does not fulfill the reader's expectations of the genre. Essentially, these authors wish students to see all medical practice as a moral enterprise, but one would suspect that they would have a hard time teaching an entire ethics course based simply on these kinds of cases. Or if such a course were taught, it would in turn be attempting to redefine the genre of the ethics case.

The bioethics discipline has been called into the medical setting to respond to a particular form of transgression and has at its central goal responding to this reportability. Loretta Kopelman, in her article discussed in chapter one, presents the following case: "Dr. T is a psychiatrist counseling Mr. and Mrs. N. They tell her that they keep their three-year-old child locked in the attic for long periods of time as punishment" (269). This two-sentence case presentation could be described as a minimalist ethics case. Recognizing what is a minimalist case in bioethics is important because it allows us to see to what degree a story has enough information in order to be considered data for ethical analysis. The first sentence of Kopelman's narration by itself could be considered a representation of a medical case, but in order for this story to become reportable as an ethics case it requires the second. If we were only given the first sentence, then we could still identify the salient moral issues of the patient-physician relationship, but the reader comes to the ethics case with the expectation not of the first sentence but of the second. The genre of the medical ethics case is, thus, a deviation from the medical case presentation. The "Basic Curricular Goals" case may possess salient moral features but it is in this second case, which deviates from the genre of the first case, that the reader's expectations for an ethics case are fulfilled. This conception of the ethics case as a break in the normal routine is akin to what John Dewey thought was the "rhythm" of our moral lives. For Dewey, our lives move normally in activities that are "confident, straightforward, organized" and then we are come upon disruption in this flow that forces us to form hypotheses as ways to resolve these problems (Pappas 108). Dewey's conception of our moral world consists thus of a series of reportable events (cf. Miller et al.).

The concept of reportability is particularly relevant for those who do work in what is usually referred to as virtue ethics. Virtue ethicists regularly note that theirs is one of the oldest paradigms for ethics, for they can trace the origin of their approach to ancient Greek philosophy. Unlike principle-based ethicists, virtue ethicists argue that we should be concerned not with generating a list of abstract principles and rules but rather with developing our moral character. That is, I should not be asking what I should do in response to a particular moral dilemma but what kind of person should I be and what a good life is. Virtue ethicists, consequently, have been highly critical of the tendency of bioethics to focus primarily on the resolution of moral dilemmas. They contend that an ethics predicated on troubling situations forces one to view ethics abstracted from the moral development of character and, in turn, leads to questions of what one should do rather than what kind of person one should be. Ethical inquiry should not be an inspection of what we should do but an examination of a way of living. Stanley Hauerwas observes, "'Problems' or 'situations' are not abstract entities that exist apart from our character" (48–49). In some ways, virtue ethicists tend to see the present concern with moral quandaries as a signal of the decline of ethics rather than, as Toulmin does, the revival of ethics. The rise of the bioethics movement can be viewed as a symptom of how ethics has become abstracted from character, the type of ethics that held sway in the West until relatively recently. The key genre is no longer the story of a life, as in the transgressive personal crisis of Augustine's *Confessions*, but the ethics case with its transgression from the medical case.

CLOSING THE CASE

In the introduction to a collection of essays on ethics consultation, John C. Fletcher, Norman Quist, and Albert R. Jonsen present a group of case examples to "illustrate the nature and range of ethical problems which arise in the clinical setting." Here is the fourth case, titled "Should Her Husband Know the Truth?":

> The child was born with cystic fibrosis, a hereditary disease which causes cysts and too much fibrous tissue in glandular organs like the pancreas and lungs. Excess mucous secretions cause a blocking of the lungs and pancreatic ducts. The disease is caused by one gene inherited from each parent, which means that both parents of a child with the disease are carriers of the cystic fibrosis gene. The mother confides in the genetic counselor who comes to see her that the biological father of the child is a man other than her husband. Yet her husband now falsely believes that he has a gene for cystic fibrosis. The counselor returns to her office and asks the physicians in the genetics program, "Should her husband know the truth?" (4)

This case possesses the reportability of many bioethics cases. Like all the case examples used in Fletcher, Quist, and Jonsen's introduction, this case ends with a question. This ending is a feature of many bioethics case narrations, and it signifies the unique closure that bioethics narratives impose on their readers; a bioethics case is a genre that requires that the reader bring closure to the plot.

Narrative, like all literature, requires closure in some form; I am using the concept of closure here in the way Barbara Herrnstein Smith defines it as, "Whether spatially or temporally perceived, a structure appears 'closed' when it is experienced as integral: coherent, complete, and stable" (2). This is the point in our experience of an artwork where we feel we have reached a sense of completion. For Smith, the work is structured to produce this experience of stability and wholeness. Because narrative discourse is so intimately tied to temporal movement, there is a tendency to associate the concept of ending with that of closure. Although in many instances this may hold true, it is not necessarily the case that Smith's concept of closure is realized in every narrative with the ending of the story. In some narrative presentations, the audience may know the ending prior to being presented with the complete narrative. Some narrative presentations consciously play with this, as in the film *Pulp Fiction*, where we see the death of one of the characters in the middle of the narrative and then see the events that led up to his death. Or take the readerly pleasure of watching a film that one has already seen; closure for fans of *Casablanca* does not come from not knowing the ending of the film. With a film like *Gandhi*, where many in the audience may know how Gandhi died prior to the beginning of the film (and for those who do not know), the film begins with his death and then shows his life up to the events that began in the narrative. Closure thus is often intimately tied to knowing the chronological ending of narrative, but can involve much more.

Narratives can differ in the degree of action required of their readers in the issue of closure. As for example a narrative, which often depends for its closure on readers rereading parts of the story a second time or closure that demands that the reader reflect on the narrative to make out the story. Bioethics cases demand that *the reader* bring closure to the narrative. The "Basic Curricular Goals" case presented at the beginning of this chapter seems odd because it neither possesses the reportability nor the closure of bioethics. Ethics cases are thus akin to what Roland Barthes refers to as a writerly text rather than a readerly one. Readerly cases are those that give readers a passive role in engaging with the text and thus it is the author who gives a narrative its closure. Writerly texts require the reader to take an active role and thus the reader is forced to write the narrative in the act of reading. In order to have closure, ethics cases require that the reader take an active role and treat the narrative in a writerly manner.

Closure in the bioethics case occurs either 1) through the lack of an ending (as the truth-telling genetics case above) or 2) through requiring that the reader rewrite the narrative.

Gustav Freytag's diagram of tragedy, which has been often extended to indicate the essential structure of all narratives (Prince, *Dictionary* 36), provides a good way to show what is expected of the reader in an ethics case: According to Freytag, stories begin with an exposition (A) followed by complication (B) and then a reversal (C) which ends with a resolution to the conflict (D).

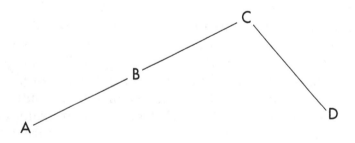

The cases presented in the Fletcher, Quist, and Jonsen work can be diagrammed as

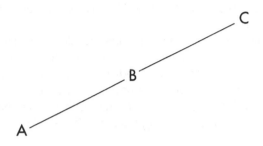

The narrative ends essentially with the climax and asks the reader to write the ending, that is, to finish the narrative.

James Childress in *Who Should Decide?* provides a perfect example of how ethicists supply cases that have A-B-C and then along with their moral argument give the reader the final ending. In his discussion of paternalism, Childress gives the reader the following case:

John K., a 32-year-old lawyer, has worried for several years about developing Huntington's chorea, a neurological disorder which appears in a person's thirties or forties, bringing rapid uncontrollable twitching and contractions and progressive, irreversible dementia, and leading to death in approximately ten years. John K.'s mother died from this disease which is autosomal dominant and afflicts 50 percent of an affected parent's offspring. Often parents have children before they are aware that one of them has the disease. John K. and his wife have a child because of contraceptive failure and an unwillingness to have an abortion because of his wife's religious convictions.

John K. has indicated to many people that he would prefer to die rather than to live and die as his mother lived and died. He is anxious, drinks heavily, and has intermittent depression, for which he sees a psychiatrist. Nevertheless, he is a productive lawyer.

He noticed facial twitching three months ago, and two

neurologists independently confirmed a diagnosis of Huntington's. He explained his situation to his psychiatrist and requested help in committing suicide. When the psychiatrist refused, John convinced him that he did not plan to act any time soon.

But when he went home, he ingested all his antidepressant medication after pinning a note to his shirt to explain his actions and to refuse any medical assistance that might be offered. His wife, whom he had not told about his diagnosis, found him unconscious and rushed him to the emergency room without removing the note. (222–223)

Childress then takes the indispensable step for any reader in ethics and that is to provide closure to the narrative by furnishing an ending. Childress stories:

When a person who has attempted suicide is brought to the emergency room, treatment should be immediate and vigorous, even if the person refuses treatment as John K. did by his note.... In most situations, the physicians and others in the emergency room will have no way to determine whether John K. was competent and wanted to die. Both care and respect demand treatment in such cases. (162)

If Childress did not complete the narrative or was not able to show how his philosophical position provided ending to this narrative, the reader would judge his position in a negative manner. In Childress's narrative, the closure is brought about in his belief that in emergency situation the ethic should be toward action. The moral position acts as the way to guide the reader toward narrative closure. The data of bioethics is thus a necessarily incomplete form, that is, it is a data that requires closure through the philosophy.

The second type of bioethics case is a narrative with an ending but an ending that does not necessarily provide closure. A narrative with an ending may still require the reader to bring closure to the narrative; in the bioethics case this may be done by requiring the reader to write it with a different, better ending. Ruth

Macklin provides an example of this in her presentation of the following case:

> Chad Green suffered from leukemia, a disease that is often fatal. His parents, believing in the efficacy of nutritional therapy and laetrile, abandoned chemotherapy, which is known to have an 80 percent rate of cure for children in Chad's age range with his type of leukemia. According to reliable medical data, Chad's "chances" were around eight out of ten for complete recovery if chemotherapy had continued and if the leukemia had remained in remission for eighteen months. Although a court ordered the Greens to resume conventional chemotherapy for Chad, they fled to Mexico to seek the treatments they believed would help. The child died some time later, and it will never be known whether he would have profited from the standard treatment. (102)

Macklin provides the reader with the end of the case but it is clear that closure has not be achieved until we decide on whether the actions of the parents should have been permitted. We must in some way rewrite the case so that either the family does not have to escape to Mexico or so that the parents were not permitted to take Chad to another country. Once again the "Basic Curricular Goals" case can be seen as one that would not normally be recognized as an ethics case for it does not require that the reader rewrite the case to demonstrate what should have happened. The rewriting gives the reader the ability to define the difference between "ought" and "is." The case presents the world as it "is" and in order for closure to occur the reader must supply a new case that presents the "ought." If the "is" and the "ought" are the same (as they are in the "Basic Curricular Goals" case) then the reader is given an essentially passive role of simply identifying salient moral components. In this we see that reportability and closure are closely aligned. In an article advocating the redefinition of the relationship of the physician and the patient, George Annas and Joseph Healey argue for the presence of a patient advocate in the hospital setting. In order to demonstrate the usefulness of this concept, the authors present a series of cases. Here is case 3:

> Mr. and Mrs. 3 have attended classes on natural childbirth. They have discussed the matter with the doctor in the out-patient clinic of the hospital where the child will be delivered. The hospital has a policy of allowing the husband in the delivery room "at the doctor's discretion." They enter the hospital and spend three hours together in the labor room. As she is being transferred to the delivery room the doctor (a resident) says to the husband, "Sorry, you can't come in, you make me nervous."
>
> In the delivery room Mrs. 3, who has previously given birth by the natural method in England, demands that the stirrups be removed. The attendants laugh at her and hold her down as her wrists are strapped to the table by leather thongs. (217)

Annas and Healey then "rewrite" the case in terms of how it would have happened differently if a patient advocate had been present.

> Under a patient advocate system, an advocate assigned to the maternity ward would be in charge of advising the medical personnel about the couple's desires concerning natural childbirth, would make whatever preparations were deemed necessary, and would be present at the parents' request to ensure during birth that the father was not denied access to the delivery room and that the mother was not subjected to coercion or ridicule—a function probably unnecessary if the husband were allowed to be present in the delivery room as a matter of course. (217)

This writerly narrative of the childbirth of Mrs. 3 permits the authors to rewrite the narrative in a manner that they believe brings closure to the reading. This construction of an alternative ending permits a satisfying closure because it creates an imagined "ought" to replace the "is": The authors fulfill a desire to take an account of the moral world and rewrite it. In Freytag's diagram, the bioethics case that provides closure in terms of rewriting can be represented in the following manner:

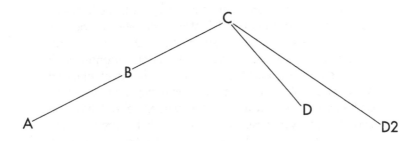

Here "D" represents the first ending, and closure is brought about by the reader offering "D2." This depends upon one acceptance that the actions of "D" were wrong and there should be an alternative ending. But I argue that if one agrees with "D" and sees no reason for "D2," then one would *not* recognize this case as an ethics case, for it already possesses closure. Of course, one could imagine other possible ways of bringing closure to a narrative, that is, there could be D3, D4, and so forth. Bioethics cases that have closure within them—as the "Basic Curricular Goals" case does—are not "genuine" ethics cases, for there is no separation of "ought" and "is." In Barthes's categories, the "Basic Curricular Goals" case is a readerly one, for the reader is not expected to provide a new version of the narrative in order to bring about closure. The bioethics case by its very nature is a writerly text rather than a readerly one. It is perhaps wrong to state—as many ethicists do—that they *solve* cases as mystery detectives do; they fulfill, instead, the reader's expectation of closure.

In *Who Lives? Who Dies?* John Kilner provides the following case:

The infant intensive care unit at Lister Hospital has reached capacity. One of the infants, John, will in all likelihood die in the next few days. Although prognosis is sometimes difficult to establish with infants less than a year old, the physician in charge and his team are virtually certain that John's massive brain hemorrhage will soon be fatal.

At this very time, an urgent call has come to the intensive care unit to admit another infant named Peter as soon as possible. Pete has sustained multiple chest and head wounds in an automobile accident, but his prognosis with intensive care is good. The physician in charge of John and of the whole infant intensive care unit is inclined to take John off the respirator, now needed for someone else. John's parents, though, would like every effort made to save their son. There are few other units in the area and all are full. A transfer to a unit in a neighboring state would put either John's or Peter's life at serious risk. (122)

What is interesting about this case is how for the reader this case lacks the part D of the Freytag diagram, the resolution. One would expect that Kilner would then furnish a possible ending and thus bring closure to the reading experience. Kilner, however, refuses to simply provide a D and instead in his comments writes a new narrative scenario in which he imagines that John's care would have been discontinued outside of the conflict concerning the competing needs of Peter.

One can imagine all sorts of emotional and legal entanglements were John to be discharged today. But these problems would be due largely to the lack of a good medical-benefit criterion, not to its presence. Were such a criterion to be clearly formulated and publicly known, parental outrage and lawsuits would be less likely—especially if John's physicians promptly discontinued intensive care when medically unwarranted rather than waiting until another arrived who needed the resources. (122–23)

Kilner's re-storying of the case demonstrates the essential writerly nature of the ethicist's data. He takes the narrative and creates another one that attempts to disturb the reportability, for Kilner's case and final closure can be diagrammed in the following manner:

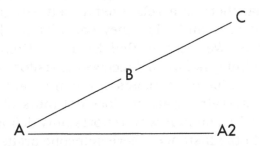

Although Kilner gives us a case where the plot would require a part D, a resolution, that links to the final crisis of C, he provides instead a rewriting that prevents the initial complication through proper medical attention. Kilner's rewriting challenges the reportability of the case as an ethics case, making it reportable for medicine but not for ethics.

As with Kilner's rewriting of the case concerning John and Peter, bioethics cases permit the reader an extraordinary degree of writerliness in the reading of the case. One can imagine a variety of different rewritings of this case, including a more traditional attempt to produce a new ending. Kilner is engaging in a particular interpretive strategy that reflects a specific argument being made in his book. He wishes to convince us of a particular way of rewriting the narrative and this entails an interpretive strategy for bringing closure to these narratives.

Narrative is so much a part of how we think about the world that we tend to forget that the world does not present itself to us with beginnings, middles, and ends. We may occasionally disagree on where a story "really begins" but we rarely argue over whether things "begin" at all. Kierkegaard observed that we live our lives forward but understand them backward, and in order to make the world understandable we create stories with distinct beginnings that then lead toward what often seems inevitable endings. I began this chapter by observing that there is a relationship between genres, such as mysteries, and what they consider the beginning of stories. We assume that Sherlock Holmes is "doing" something before he receives a visitor asking for help, that is, his life consists of the mundane activities, but his narrator, Watson, does not think of these as reportable. In a similar way, reportability defines the ethics case. In the instance where someone argues that the case just presented "is not an ethics case," it seems what is at stake in this claim is more than just classifying a narrative but in defining what is "ethics." We read ethics cases, however, as we do mysteries, not only for their openings but also for their endings. Mysteries invite their readers to bring closure to the narrative with the detective, but bioethics cases leave narrative in some unfinished state. Some bioethics cases are left literally unfinished and thereby require that the reader or the philosophy finish the narrative. Other cases demand closure by having one rewrite the narratives. Moral paradigms in bioethics such as principlism, casuistry, and care ethics are manuals for bringing closure to our moral problems. In the end, bioethics is a battle over the tock of the case.

Chapter Six: Why Bioethics Lacks Character

> "But isn't a person, and, to an even greater extent, a character in a novel, by definition a unique, inimitable being?"
>
> —*Milan Kundera*

Tzvetan Todorov divides narratives into those that focus on character and those that focus on plot. He describes these as psychological versus apsychological narratives. Apsychological narratives are those in which "the actions are not there to 'illustrate' character but in which, on the contrary, the characters are subservient to the action; where, moreover, the word 'character' signifies something altogether different from psychological coherence or the description of idiosyncrasy" (*Poetics* 66). Psychological narratives derive their aesthetic continuity from the exploration of character. Henry James thought such a division was a false dichotomy, in that it is impossible to separate character from action: "What is character but the determination of incident? What is incident but the illustration of character?" (18).

Although I am sympathetic to James's point, I

believe that Todorov's division can be modified to a continuum, with action on one side and character on the other, rather than a simple dichotomy of either one or the other. There are genres that tend toward the psychological and others that tend toward the apsychological. Some genres are clearly within the confines of apsychological narratives in that one reads the story primarily for its action and not for character. Adventure narratives are possibly the clearest example of this. In the film versions of the James Bond series different actors can play the same role and not lose anything substantial from the main focus of the narratives. The other side of the continuum can be illustrated by such writers as Virginia Woolf where it is the dimension of depth afforded to the characters that is the main objective of the narrative. In her *To The Lighthouse*, very little happens in the way of plot. The narrative hangs directly upon the development of character, and the entire narrative would be different with different characters. Other genres seem to have elements that contain both the psychological and the apsychological. The detective genre, which emphasizes action over character analysis, can develop a psychological element because of the continuity of different figures. Characters such as Tony Hillerman's police officer Jim Chee or the TV character Frank Pembleton in *Homicide* both become psychological studies; the receiver (i.e., the reader or the viewer) observes these characters struggle with personal relationships and religious beliefs over a series of what tend to be apsychological crime plots.

The ethics case clearly falls within the continuum on the side of the apsychological narrative. This may at first appear strange considering that decisions about human life and death would seem to demand such a depth, but characters in ethics cases are often interchangeable within the drama. In Albert Jonsen, Mark Siegler, and William Winslade's textbook,

Clinical Ethics, four clinical cases are used as paradigms for various types of medical conditions. "The
patients in these cases are given the names Mr. Cure,
Mrs. Cope, Mrs. Care, and Ms. Comfort. These fictional names are chosen to suggest certain prominent features of their medical condition" (10). These are not so
much representations of people as they are of medical
conditions; as in medicine itself, people are clearly seen
as instances of disease.

This emphasis can be seen in the way ethics cases
"present" characters. Like other apsychological genres,
the bioethics case presents character traits only when
they directly relate to the cause of actions. For instance,
in a case included in Tom Beauchamp and James
Childress's classic textbook, *Principles of Biomedical
Ethics*, the authors tell of a man who has cancer but is
not told his full diagnosis. After describing his treatment, the narrator states, "The patient, a hard-driving
entrepreneur who dominated both his family and his
business, was first told that there was a probability of
'malignant cell transformation' in his thyroid" (3rd ed.
405). In the narrative, this characterization then
results in the man going back to work "against medical
advice—until he exhausted himself." Like many examples in the genre of the ethics case, this description
becomes predictive of later action, and no characterizations are provided that do not immediately relate to
the character's action. Characterization thus is only
provided when it in turn furnishes a degree of plausi-
bility to the action. If, in the above case, the reader were
not told of the man's tendency to work hard, then his
actions might seem odd and might signal to the reader
that he was actually aware of his diagnosis. Instead, the
character trait becomes directly related to the action
and would not have been told if it were not essential to
driving the plot forward. In Roland Barthes's dissection
of the various codes within a narrative, he distin-

guished the proairetic code from the semic code. The proairetic code concerns the sphere of actions within a text; the semic concerns the various attributes that constitute a "character." In the ethics case—as in the "hard-driving entrepreneur"—the semic code always exists to support the proairetic code, and thus is in reality simply a subcode within the proairetic. It is for this reason that in the genre of the ethics case there are not so much "characters" as there are "roles."

WHAT'S IN A NAME?

Another way characters are represented in ethics cases is through the designation of particular names. For example, one character appears again and again in E. Haavi Morreim's analysis of medical ethics and economics, *Balancing Act*: the mysterious "Mrs. Jones." In chapter one, Morreim introduces her to the reader, as if we were expected to already know her and her family. After claiming that the moral lives we lead are not comprised "of terrible hypotheticals from which there is no escape," Morreim argues:

> Under ordinary circumstances one's moral aim is rather to forge some resolution that will maximally honor all the important competing values. If the impoverished Jones family vows that they cannot afford to have another child they do not, upon discovering that Mrs. Jones is again pregnant, ask themselves "Gosh, which of the children shall we throw out onto the streets?" (5)

Mrs. Jones is not only pregnant, but we learn in chapter three that she is in the hospital with an intravenous line. In chapter four, Mrs. Jones is recovering from pneumonia. And in chapter six Mrs. Jones needs a mammogram and later elective surgery. Who is Mrs. Jones? Why are we expected to know her? Of course, we both do and do not know Mrs. Jones. In a note at the beginning of her book, Morreim states that, "all identifiable information used in case examples has been changed, and the structure of the case example altered, so as not to identify the individuals involved" (xiii). The name "Mrs. Jones" acts as a marker for these other people whom Morreim wishes to protect. David Lodge observes that names we give characters "are never neutral. They always signify, if it is only ordinariness" (37). Most probably Morreim did not intend this marker to indicate a person with many medical issues but rather to indicate that these are the kind of medical and social issues anyone may face. Mrs. Jones is not always alone, sometimes she sits within a web of other names, which also indicates the ordinariness of these problems. For instance, Mrs. Jones's pneumonia is actually portrayed in relation to a group of individuals: "If Mrs. Baker is admitted to the lone available intensive care bed, then Mr. Abel, also in need, is not. But Mrs. Jones, recovering from pneumonia on another ward, is unaffected" (50).

But for whom are these names neutral and ordinary? The idea of neutral names relies upon a social context in which those names signify "ordinariness." Barthes observes that the proper name of characters contributes a cultural trait that adds to their other character traits (S/Z). So when Morreim gives her character the proper name "Mrs. Jones" it brings with it associations of the common and the ordinary. Yet for whom is Mrs. Jones ordinary? There is nothing inherently ordinary or common with the proper name "Jones," except when placed within the Anglo-American context. Jones, Baker, and Abel are signifiers of the average and ordinary person within this particular social context and would lose this meaning if the cases were written for a Mexican or Israeli audience and become markers for "any North American." Consequently, Mrs. Jones both is and is not ordinary, depending on the norms of a particular community of readers. The significance of this lies in the degree to which cases are written with an assumed image of the average patient. Ironically, Mrs. Jones becomes unusual or rather significant primarily because she is thought by the writer to be ordinary and common. Davis recognizes this issue when she notes how she was preparing to hide the identity of some patients in several ethics cases ("rather enjoying the prospect of playing with the details, inventing some names, adding a little embroidery, fulfilling my novelistic fantasies") and soon realized that changing these features radically altered the very features she wished to analyze. Although Davis is not particularly concerned with the issue of proper names in altering the features of the case, one can imagine how a case involving racism in the treatment of a Mexican-American could not be altered by changing the name to Jones or Smith.

The cultural traces that follow the choice of a particular proper name cannot be neutralized by employing what might at first seem to be empty signifiers. When bioethicists use initials, for example, instead of creating a name, as Terrence Ackerman and Carson Strong do in *A Casebook of Medical Ethics*, they are still drawing upon a fiction used in medical case presentations. Or look, for example, at a case presented by Loretta Kopelman, which begins in the following way: "Miss M is 21 and has been seeing Dr. Z as a psychiatric outpatient" (271). Although both patient and caretaker are designated through single initials, the additions of "Miss" and "Dr." carry with them power and gender relations within contemporary Western society. Similarly, some cases attempt to refer to characters purely through their particular roles, e.g., doctor, social worker, or patient. This common feature actually makes a great deal of sense when viewed in relation to the tendency of the ethics case to be action-driven— that is, for characters to be defined primarily by their roles rather than by an intrinsic quality.

THE ACTANTIAL MODEL AND THE ETHICS CASE

A major feature of structuralist approaches in literary criticism has been to analyze characters in terms of roles rather than as entities understood outside of the plot. The insights of this approach provide some provocative entrance points for examining the bioethics case. In part, structuralists were interested in determining the fundamental structure of all narratives. Vladimir Propp was the first to break narrative into a series of constituent parts. In *Morphology of the Folktale*, he suggested that there are certain deep structures that exist within all folktales. In a move that is similar to that of the linguist Ferdinand de Saussure, Propp argued that one should not study the various types of folktales but instead attempt to discern the structural system that all these narratives share. He observed that the Russian folktales he analyzed shared two common features. First, a constant series of thirty-two *functions* appear in the plot of the folktale in a particular sequential order. Second, and more relevant to our discussion here of character, there are *roles* in all of the folktales, parts which can be filled by a variety of characters but are intimately tied to their task within the structure of the plot. These roles are:

the dispatcher

the sought-for person

the hero

the donor

the villain

the helper

the false hero

Although Propp made no claims for this system outside his analysis of the Russian folktale, he supplied
structuralism with some essential forms through
which to view characters in terms of their part within
discourse. As Barthes observes, "Anxious not to
define character in terms of psychological essences,
structural analysis so far attempted ... to define the
character as a 'participant' rather than a 'being'" (as
quoted in Culler 232). The subsequent work of nar-
ratologists drew upon and attempted to refine Propp's
initial insights.

A. J. Greimas, one of those scholars who Robert
Scholes refers to as the "progeny of Propp," has
attempted to refine and generalize Propp's original set
of roles. Greimas's intent was to construct explicitly a
set of relationships that represented the essence of a
narrative. He argued that narrative consisted of six
actants, which draw upon yet structurally clarify and
improve Propp's original set of roles:

Subject

Object

Sender

Receiver

Helper

Opponent

In a story a sender sends a subject to find an object for a receiver, and the subject's efforts are either helped or opposed or both in sequence. A single agent can take the role of several actants, and actants need not be persons but can be inanimate objects as well. Greimas's work is strongly influenced by structural linguistics, and this is evident in how the meaning of each actant is dependent upon their relation to their difference from other actants. Just as a word in a language gains its meaning through its distinction from other words (man/woman; up/down; inside/outside) so the meanings of the actants within a narrative exist in opposition to other actants, and thus Greimas saw the actants as a series of interdependent oppositions.

Subject	Sender	Helper
Object	Receiver	Opponent

In terms of this model, one should be able to take any narrative and strip down its agents into a series of actantial relationships. So a narrative such as "Little Red Riding Hood" could be analyzed in the following manner:

Subject	Little Red Riding Hood
Object	Grandma

Sender	Grandma's Sickness
Receiver	Grandma

Helper	Woodsman
Opponent	Wolf

Yet one can use this model of story to break down even more complicated pieces of literature. We can use it to analyze narratives such as F. Scott Fitzgerald's *The Great Gatsby*:

Subject	Gatsby
Object	Daisy

Sender	Love
Receiver	Gatsby

Helper	Nick Carraway
Opponent	Daisy's husband

For Greimas, each of these pairings results in different ways of classifying narratives. For example, one can group those narratives in which two of the actants are the same character; in folktales, like "Little Red Riding Hood," and modern novels, like *The Great Gatsby*, the subject and receiver are the same characters. Another possible distinction, according to Jonathan Culler, "is between stories in which helper and opponent are separate characters and those in which they are fused in one or more ambivalent characters" (234). Shakespeare's *Hamlet*, for example, is a narrative in which it is difficult to ascertain whether characters are helpers or opponents.

A return to the "Basic Curricular Goals" case discussed already can be instructive. This case concerned the typical medical story of a patient with an illness who comes to a physician for help; the patient is informed of the various treatments for the illness and after consenting to the treatment recovers. We can parse this narrative in the following manner:

Subject	Patient
Object	Health
Sender	Disease
Receiver	Patient
Helper	Physician
Opponent	Disease

This narrative of a medical encounter represents the physician's essential moral role as the patient's helper in attaining the goal of health. Respecting the patient's autonomy is essential in determining the patient's object and helping the subject reach that object. As is the ideal of the medical encounter, this case has the physician as helper, the patient as subject and receiver.

The reportability of an ethics case arises when the question of the physician's position as helper becomes of major concern. In other words, it is an example of Culler's observation of ambiguity in the roles of helper and opponent, and thus it is this actantial opposition of physician as helper/opponent that is central in understanding the reportability and the closure required of the ethics case. If we alter features of the "Basic Curricular Goals" case so that the obvious malady has no effective cure and the patient's object is a "dignified death," requiring (from the patient's perspective) assistance in committing suicide, the physician's role in the narrative turns on the decision whether to be a helper or an opponent. The case is not a moral problem until the physician must decide what role should be taken and, as with physician-assisted suicide, whether certain actions are not viewed by society as being a helper. Compare the binary splits of the "Basic Curricular Goals" case to the following case on patient-physician confidentiality from Beauchamp and Childress's *Principles of Biomedical Ethics*:

> After experiencing dry, persistent coughing for several weeks and night sweats for ten days, a bisexual male visits his family physician. When the patient describes his symptoms and admits that he is bisexual, the physician orders a test to determine if the patient has antibodies to the human immunodeficiency virus (HIV), the virus that causes AIDS. The test results are positive and indicate that the patient has been infected with the virus, will probably develop full-blown AIDS over time, will probably die from the disease, and is probably capable of infecting others through sexual contact. In a long counseling session, the physician explains all this to the patient and discusses the risk of unprotected sexual intercourse to his wife, as well as the possibility that their children, now one and three years old, would be left without parents if his wife contracts the disease too. The patient refuses to allow the physician to disclose his condition to his wife. The physician finally and reluctantly accedes to this demand for absolute confidentiality. After surviving two episodes of opportunistic infection, the patient dies eighteen months later. Only during the last few weeks of his life does he allow his wife to be informed that he has AIDS. She is then tested and is found to be antibody positive, but she does not yet have any symptoms. However, a year later she goes to the doctor

with dry cough, fever, and loss of appetite. She angrily accuses the physician of violating his moral responsibility to her and her children; she insists that she might have been able to take steps to reduce the risk to herself if she had only known the truth. (403–404)

This case reveals how this ambiguity of the helper/opponent opposition plays such a central role in the bioethics case:

Subject	Patient
Object	Secrecy
Sender	Disease
Receiver	Patient
Helper	Physician
Opponent	

One should take note that I have left out the position of the opponent, which some may wish to see as the wife. Yet the issue of the opponent is a key one in the bioethics case. The writerly closure demanded by the reader to decide whether the physician should be a helper or opponent reflects what H. Tristram Engelhardt, Jr. refers to as the "conflict at the roots of bioethics" (66), the split between beneficence and respect for autonomy. If the physician decides to "oppose" the patient in his desire to maintain a secret from the family, then the physician is acting to do what he believes is the beneficial action to the family. When the physician follows the object of the patient's desire, then he is helping him reach an object that may be in opposition to the traditional goals of medicine.

This actantial analysis that has been developed by structuralists has most often been tested upon folktales, myths, and adventure novels, which, like the bioethics case, are plot-driven and apsychological, and consequently the insights of structuralists are particularly relevant to understanding the genre of the bioethics case. The serviceability of the structuralist model reveals how cases reflect principlist concepts. Principlism has arisen from a belief that there are certain prima facie moral principles, and moral problems arise in medicine when there is a conflict between particular principles. For many who follow this paradigm, the way to "work up" a moral problem is by first determining the principles in conflict and then determining which should take priority in the particular case. What makes the structuralist agenda easily applicable to particular forms of genres—as for example Barthes's analysis of the James Bond novels—demonstrates the limitations of the data of bioethics and its tendencies to be driven in terms of conflicting principles. By contrast, the structuralist approach is inadequate in analyzing Joycean-style narratives, which are far more psychological. Thus, an ethics case written in a psychological manner—one that breaks from the genre's conventions—would make a simple binary split between principles seem deficient to some degree.

POINT OF VIEW AND THE DESIGNATION OF THE SUBJECT

Greimas is known for continually "refining" his original set of actantial relationships. In some of his later models he added the anti-subject (Greimas and Courtés). The anti-subject is not simply trying to prevent the subject from reaching the object but is doing so because the anti-subject is attempting to reach an object as well; the two are rivals. The object could of course be the same object of the subject, as in folktales where both the hero and the villain of the story are trying to win the treasure. A return to the AIDS confidentiality case illustrates the complexity of this type of analysis, for one could formulate the analysis in the following manner:

Subject	Husband
————	————
Object	Secrecy

Sender	Disease
————	————
Receiver	Husband

Anti-Subject	Wife
————	————
Object	Health

Yet as Scholes notes, Greimas's distinction of subject and object in a narrative can be confusing for "subject and object are matters of point-of-view and are therefore reversible. In a love story ... Boy and Girl may both be Subject and Object, both be Sender and Receiver" (105). Scholes sees this as an unnecessary confusion, which Etienne Souriau's earlier model (upon which Greimas also drew) designated as "the one who desires." Yet I believe Scholes's insight brings forth questions of how interdependent the subject of a narrative is with the point of view designated in the story, and consequently we may not be able to simply construct an actantial analysis without in essence referring to the point of view constructed by the author. If the case were told from the wife's point of view she would be the subject and her husband would be the anti-subject.

Subject	Wife
————	————
Object	Health

Sender	Disease
Receiver	Wife, Children

Helper	
Opponent	Physician, Patient

This type of switch actually occurs in the narrative for it begins with the narrative told with husband as subject and the wife as anti-subject, but following the husband's death the wife becomes the subject of the narrative with the physician as the anti-subject.

We, thus, return again to the importance of point of view in the construction of moral positions. At the beginning of this chapter, I introduced Todorov's distinction between psychological and apsychological narratives and I argued that ethics cases tend to be driven by plot rather than character. Morreim's use of "generic" names indicates the degree in which characters are not so much particular individuals as representatives of particular situations. This lack of interest in the particularity of characters in bioethics results in moral paradigms that focus on this very issue, such as care ethics and phenomenology. They are not given the data they need in order to make their analysis seem relevant. The tendency of ethics cases to be action-oriented leads the reader to see moral problems through binary choices. How we perceive cases is influenced, however, by whose point of view the narrative is presented. Who are the subjects and the anti-subjects of a case? Such choices bias how readers understand the moral issues of a case. Shifts in points of view can alter the values readers see as in conflict. It seems that the author of ethics cases has a moral duty to give voice to as many perspectives as possible. For this reason, we should now turn to how the voice of the patient is represented in the case.

Chapter Seven: Speaking for the Patient

In Patricia Highsmith's classic crime novel, *The Talented Mr. Ripley*, the main character temporarily assumes the identity of a man whom he has murdered in Italy. At the end of the novel, Ripley, who has returned to his original identity, goes to get his mail and fears that his plot has been uncovered.

> "Would you stop at the American Express, please?" he asked the driver in Italian, but the driver apparently understood "American Express" at least, and drove off. Tom remembered when he had said the same words to the taxi driver in Rome, the day he had been on his way to Palermo. How sure of himself he'd been that day, just after he had given Marge the slip at the Inghilterra!
>
> He sat up when he saw the American Express sign, and looked around the building for policemen. Perhaps the police were inside. In Italian, he asked the driver to wait, and the driver seemed to understand this too, and touched his cap. There was a specious ease about everything, like the moment just before something was going to explode. Tom looked around inside the American Express lobby. Nothing unusual. Maybe the minute he mentioned his name—
>
> "Have you any letters for Thomas Ripley?" he asked in a low voice in English.

"Reepley? Spell it, if you please."
He spelt it.
She turned and got some letters from a cubbyhole.
Nothing was happening.
"Three letters," she said in English, smiling. (403)

In just this brief section, Highsmith presents the discourse in an extremely complex manner. The author draws upon conventions so familiar to readers of modern fiction that the fact that Ripley's words and thoughts are presented in such an elaborate manner is most probably not immediately obvious. First, Ripley speaks to the driver but there are no direct quotations. Then we have what appears to be Ripley's thoughts: "Nothing unusual. Maybe the minute he mentioned his name," and then we have the words exchanged between Tom and the woman at the American Express office.

Gérard Genette distinguishes three basic ways that discourse can be represented in narration: reported, transformed, and narratized. *Reported* discourse, also known as "direct discourse," is the most easily identified in a narrative, for it represents the "exact" words spoken by a character. The Highsmith quotation above is framed by two reported speech sentences. It begins with "'Would you stop at the American Express, please?' he asked" and ends with "'Three letters,' she said in English, smiling." The fact that I have to provide double quotation marks to indicate the difference between the narrator's speech and the character's speech illustrates the clearly divided language acts in this form of discourse. Both sentences also include "tag clauses," which identify the speakers and furthermore characterize the tone of the discourses presented.

In *transposed* discourse the character's words are not presented in the exact fashion they are said or thought but instead are summarized or re-represented by the narrator. The sentence: "In Italian, he asked the driver to wait" is an example of transformed discourse

114

but the transformed discourse can also be presented in terms of the character's thoughts, as in the sentences "Tom remembered when he had said the same words to the taxi driver in Rome.... How sure of himself he'd been that day, just after he had given Marge the slip at the Inghilterra!" One of the notable elements of transformed discourse is that the first-person becomes the third-person and consequently it represents language liminally situated between the narrator and the character.

Narratized discourse, as Gerald Prince defines it, is "a discourse about words uttered (or thoughts) equivalent to a discourse not about words" (*Dictionary* 64). In the Highsmith quotation, this can be seen in the sentence "He spelt it." This is not reported discourse nor is it transformed, but instead it reports that discourse has happened just as the sentence above reports, "Tom looked around inside the American Express lobby."

Except for free direct discourse, when bioethicists portray moral problems, they employ all of these forms. They tend, however, to use narratized and transformed discourse. The reason for this lies in the degree to which the narrator is able to "control" the speech of others. As I will show below, we can judge the freedom that an author permits characters by the way the author manipulates their language. Since bioethics promotes such moral principles as patient autonomy and self-determination, we may expect the discipline to attend to the degree by which patients have control over the representation of their ideas and feelings.

TRANSFORMED DISCOURSE AND REPORTED FRAGMENTS

Look at the representation of discourse in the following ethics case presentation from Baruch Brody and H. Tristram Engelhardt, Jr.'s textbook *Bioethics*:

Mrs. A, a 35-year-old secretary, is found after her routine

pap smear to have carcinoma of the cervix. The cancer is in an early stage, confined to the cervix, with minor invasion (microinvasion) of the tissue of the cervix (stage 1A). The tumor is still easily treatable by means of a simple hysterectomy. Even though the physician informs Mrs. A that there is a 90 percent chance of a complete cure, she is very distraught and apprehensive. When the physician raises the issue of scheduling a time for surgery, Mrs. A is evasive and says she want to think things over. The physician makes another appointment to speak with her a week later. Mrs. A still is very reluctant to discuss the matter of surgery but agrees that "most likely, it is something I will have to face." She then asks what risks would be involved in a hysterectomy. The physician is concerned that if he describes all of the risks in full detail (for example, injury to the bladder, to the urethra [the tube that connects the kidney and the bladder], injury to the bowel and/or intestinal obstruction, and urinary incontinence, not to mention possible death due to anesthesia), the woman will postpone the surgery even further. He wonders whether he should instead inform the woman's husband of the usual risks and indicate to the woman more globally that things will likely go well and that people are able to return home from the hospital after only a few days. (285)

Brody and Engelhardt rely primarily on transformed and narratized discourses. For instance, the narrator first provides an instance of transformed discourse in, "When the physician raises the issue of scheduling a time for surgery, Mrs. A is evasive and says she want to think things over," and then the next sentence is an instance of narratized discourse: "The physician makes another appointment to speak with her a week later."

One sentence does include an instance of reported discourse, but it is introduced by transformed dialogue: "Mrs. A still is very reluctant to discuss the matter of surgery but agrees that 'most likely, it is something I will have to face.'" This pattern illustrates a common feature of discourse within bioethics. Pieces of reported discourse are inserted within longer pieces of transformed discourse as if bioethics authors wished to include common phrases that patients characteristically state, so one finds, for instance, such fragments of reported discourse as "the patient reported repeatedly that she 'didn't want to be hooked up to those damn machines.'" These phrases are used to signify to the reader that the cases come from "real" events that can only be quoted in bits and pieces of conversations remembered. These reported fragments, however, raise issues of the degree to which patients are permitted to speak for themselves. These fragments within transformed discourse unintentionally signify the control the author possesses over what is considered "relevant" information. Michael Toolan observes, concerning transformed discourse, that "[s]uch narratorial speech-summarizing is useful in reporting unimportant conversation, where a verbatim account seems aesthetically undesirable—or for referring a second time to a conversation that has been previously presented to the reader more fully. It is accordingly quite widespread in novels" (122). For the discipline of bioethics, however, one should ask who is determining what is "unimportant" conversation.

THE DRAMA OF REPORTED DISCOURSE

It may seem at first that reported discourse is more "genuine" than narratized and transformed discourse, but this conclusion only privileges one form of representation over others. As Genette reminds us, "mimesis in words can only be mimesis of words" (*Narrative* 164), so we must remain cognizant that even representation of speech as spoken, as in reported discourse, is still a representation. Even in reported discourse, "there is a narrator who 'quotes' the characters' speech, thus reducing the directness of 'showing'" (Rimmon-Kenan 108).

Reported discourse in bioethics cases often signals a moment of high drama, the moment of crisis. Look at the following case presented from Ernlé W. D. Young's *Alpha and Omega: Ethics at the Frontiers of Life and Death*:

> A twenty-three-year-old man was found one night, stabbed through the heart, on a street in East Palo Alto. He was immediately rushed to the emergency department at Stanford University Hospital. From there he was taken into the operating room for heart surgery. The surgery was successful. The young man's heart was repaired. But after the operation he did not regain consciousness. Subsequent neurological tests revealed that he had sustained irreversible and almost total brain damage during the time he had lain bleeding on the street where he had been found. Instead of being pumped to his brain, his blood had flowed out into the gutter.
>
> A month later, the physicians taking care of him came to the unanimous conclusion that this young man's life should no longer be sustained by artificial means and that he should be allowed to die naturally. The patient's father, who had lost his only other child in a drug-related violent incident seven months previously, refused to listen to any suggestion that his surviving son not now be treated as vigorously as possible. When told that the victim could lie in a vegetative state for the next forty years and would probably never again communicate meaningfully with anyone, his impassioned response was, "Even if I have to lead him around on a leash like a dog for the rest of his life, I want my boy alive, not dead." For several days, a stalemate ensued, with the father refusing all requests by the medical team to give them permission to stop treating the patient aggressively and to start caring for him palliatively. And then the problem was unexpectedly resolved when the young man went into cardiac arrest and died before he could be resuscitated. (27)

In this case the narrator provides narratized and transformed discourse, and, thus, it is typical of case presentations in bioethics. The drama of the case develops, however, at the moment that the father "speaks" directly in the case. In numerous ethics cases, the moment of crisis occurs when the discourse suddenly changes from transformed to reported discourse. The effect on the reader is that the moral drama occurs as soon as the patient's voice "breaks through" into the ongoing medical care.

The following case from Mark Waymack and George Taler's *Medical Ethics and the Elderly* provides another example of this drama of reported discourse:

> The surgeon finally appeared at the doorway, hesitated, and then stepped into the waiting room with the bad news: mother's biopsies were positive. The tumor had obviously spread.
>
> Mr. Castle looked resignedly at his wife, who obviously shared the sense of grief for Mrs. Castle as a mother-in-law and as another woman. Their daughter, Lisa, began to cry, and her parents moved away from the surgeon to comfort her.
>
> Later, the family met with Mrs. Castle, who stated, "I want to go home. I want to die in my own bed, like my mother did. Laura, you *will* take care of me, won't you?"
>
> The elder Mrs. Castle instinctively knew that this was her last illness and prayed that her end would come swiftly. She also knew that she was in competition for attention with her spiteful granddaughter, of whom she was already quite resentful. The girl was not having an easy time of adolescence, sampling drugs, acting rebelliously, and doing poorly in school. She was now remarkably subdued, restrained like an angry cat, sitting coiled in the corner of the hospital room.
>
> The couple turned to the surgeon who, despite his initial demeanor, seemed very hopeful that chemotherapy and radiation would give her another couple of good years, "maybe three to five." He was like a judge passing sentence, glad that a difficult trial was approaching an end. They, however, saw only a prolonged course of recurrence, progressive disability, and eventually death. A tough road to travel in any case, but with problems with Lisa, it seemed overwhelming.
>
> "We think," ventured Mr. Castle, "that mother would be better off in a nursing home. We'll visit often ... but we just can't handle her at home." (52–53)

Here we have two instances of reported discourse that create the drama of the case. There is the reported fragment of the surgeon's "'maybe three to five.'" There is also a notable amount of transformed discourse, especially in the thoughts of the parents: "They, however, saw only a prolonged course of recurrence, progressive disability, and eventually death. A tough road to travel in any case, but with problems with Lisa, it seemed overwhelming." The moment when Mrs. Castle makes the statement that she wants to "go home" and Mr. Castle explains to the surgeon that "'we just can't handle her at home'" the case becomes a problem for the surgeon. Ironically, the case would not have been an ethics case if the various participants had not made such statements.

In the earlier discussion of duration in case narrations, I gave an example of a case by John Lantos that recreated the time the action occurred in the reading time of the narrative. Look at the following selection:

"How can an 18-year-old make a decision like that?"
"According to the law, she can make it."
"According to your law. Not according to her law."

The case's ability to provide a parallel between the narrated time and the narrative time is due to the use of reported discourse, but (this case was part of a special case series that strove to provide alternative forms of case presentations [Miles and Hunter]) such cases are rare in bioethics.

THE PHENOMENOLOGY OF FREE INDIRECT DISCOURSE

Consider, then, the case of a 74-year-old woman, widowed for over ten years. She has two sons, both of whom are married with their own families. One son persuaded her to live in the same small town as he; the other son lives a long distance away. Some years earlier, after the death of her husband, the woman underwent serious surgery, which left her debilitated for some months. No sooner had she begun to recover enough to be up and about, however, than she began to develop serious arthritis—first in her hands and legs, then over the next seven years throughout her body. As she is reported to have said more than once: "I hurt in every joint I've got and in some I never knew I had." She tried every sort of treatment and medication available, went to specialist after specialist, but to no avail. The treatments either caused severe stomach pains ("they all make my stomach act up somethin' awful!") or other debilitation side effects (the gold salts caused extensive shingles lasting over a year, for instance).

This is the first paragraph of a case titled "Why Won't You Let Me Die?" from Richard Zaner's *Ethics and the Clinical Encounter*. Zaner strives to present this case "in considerable depth and detail." The story concerns a woman whose health steadily declines to the point where she is placed on a ventilator. After she recovers, she writes advanced directives concerning her wish that physicians not use any mechanical methods to postpone her death. Later she is found comatose and placed on a ventilator, and her sons struggle with the attending physician to remove the artificial measures keeping her alive. The sons wish to follow their mother's wishes; the physician is unwilling to discontinue life support until he is confident that she is incurable. Eventually the ventilator is removed, and the woman dies.

My summary does an injustice to Zaner's extraordinarily detailed account. The length of Zaner's case presentation is one of the remarkable features of his style. Though his style has many noteworthy aspects, I wish to restrict my comments in this analysis to how he uses dialogue.

In this first paragraph (presented above) Zaner

quotes the woman's subjective account of her arthritis in moments of reported discourse: "I hurt in every joint I've got and in some I never knew I had." This quotation not only offers the "subjective" perspective to balance the physician's "objective" view of her condition, it also characterizes the woman, for before the narrator expressly describes her character, the cadence and tone of her speech reveal to the reader what kind of woman this is. Her use of the word "got" and the idiomatic phrase, "and in some I never knew I had" gives us an image of a woman who is down-to-earth, folksy, and perhaps lacking formal education. This characterization is reinforced by the second quote: "'they all make my stomach act up somethin' awful!'" She refers to her stomach pains as the medication making her stomach "act up" and then the narrator conveys her diction by dropping the "g" in "somethin' awful." This woman's style of speech is in distinct contrast to the narrator's statement of her situation just before: "The treatments ... caused severe stomach pains."

Yet throughout his narration of this case, the narrator never fully quotes the people within the narrative beyond a single sentence. Rarely is a complete sentence given, but the reader is provided instead with reported fragments woven in the narrator's own statements:

> She had always said to her sons that she would never want to live if she could no longer "do for myself." Fiercely independent, prizing being "beholden to nobody," she now found herself facing what was for her the worst of all possible fates: a nursing home, where she would have few of "my things," little time for herself on her own terms, where everything and everyone is "run" on a schedule dictated by the needs of the home itself and not by her needs and desires.

Often the narrator interjects comments or edits his characters's dialogue. Take, for example, his representation of the first son's difficulty in getting his mother off the machines. "Besides, since these doctors were not able (or willing) to say whether her death was 'imminent,' they had to run their tests, *didn't they?*" This is a remarkable sentence in its weaving together of quotations and descriptions, its entanglement of Zaner's thoughts and the son's. Who is the person making this statement? Is it Zaner or the son? In this passage Zaner employs a particular form of transformed narration—free indirect discourse. If the discourse of the son were represented in reported discourse, it might look like the following: "The son said, 'Besides, since these doctors were not able to say whether her death was "imminent," they had to run their tests, didn't they?'" In indirect discourse, a form of transformed discourse, the statement could be represented in the following way: "The son said that the doctors were not able to say whether her death was "imminent," and so didn't they have to run their tests?" Free indirect discourse consists linguistically of a mixture of the features of direct discourse and indirect discourse (see McHale). Dorrit Cohn contends that free indirect discourse, "may be most succinctly defined as the technique for rendering a character's thought in their own idiom, while maintaining the third-person references and the basic tense of narration" (100). Others have pointed out that free indirect discourse represents a form of speech "transformed" primarily through the fusion of the narrator's and the character's voice. This fusion is referred to as the "dual voice." Ann Banfield summarizes this position on free indirect discourse as one that "denies the distinctness of sentences of pure narration and those of represented speech and thought, offering as its primary evidence ... a third kind of narrative sentence where narrator's and character's points of view are said to 'merge'" (185). Zaner's narrator's continual use of free indirect discourse fuses the two speakers together. Zaner's description is "thick" not because the narrator has an omniscient view of the situation—which the narrator does—but because the narrator is omnipresent, or perhaps we should say omnivocal. The narrator is in everyone's speech, continually speaking with or for them. The narrator characterizes the speech before people say it and seldom allows characters to speak for themselves. Take the encounter that the second son has with a nurse after speaking with Dr. Jones:

> It was as Jones had told his brother, but she went on to say—with words designed as much to elicit his feelings and response as to describe what she knew—that she had noticed what may have been some slight spontaneous leg movements, earlier that afternoon as she was turning his mother's body.

Here again the narrator interjects comments on the purpose of the nurse's words before telling us what she said. Or take the following example of the second son's confrontation with Dr. Jones:

> Jones's response was swift and harsh. Ethics, he asserted, was irrelevant here and now, as the issue was strictly medical and legal…. "But, as legally binding, the thing says you've got to comply or else transfer her to a physician who will comply?" "No way," Jones replied, not unless and until he was himself clear that her medical condition was incurable and terminal even with life-supports in place.

There is little separation in Zaner's work between the narrator's voice and other people's voice, for his narrator moves seamlessly between reported and transformed speech. "'No way,' Jones replied, not unless and until he was himself clear that her medical condition was incurable and terminal even with life-supports in place." Here he proceeds quickly from directly reporting Jones's speech to indirectly describing the rest of Jones's contention. Throughout Zaner's description of the events, his style of quotation demonstrates the ethicist's presence in the narrative. The narrator comes in to describe and define the conditions of other people's speech and the direction of their comments.

What is the significance of Zaner's style? I believe we must look at his methodological bias in the discussion of bioethics. Zaner is critical of the standard method of bringing forth a multitude of cases to illustrate a principle, and he advocates a phenomenological approach to the resolution of ethical problems, one that begins with a close study of one particular ethical dilemma. Howard Brody has characterized Zaner's methodology as "describing a case with sufficient richness of detail and sufficient understanding of the views and motives of all involved parties so that an appropriate resolution will emerge from the telling" (*Healer's* 250). Zaner consciously includes more detail in his presentation of the social facts than most ethicists, but the accumulation of details does not designate an approach as phenomenological. As Zaner observes in his commentary on "Why Won't You Let Me Die?," "Every clinical case requires the same kinds of paralinguistic, conversational, and contextual probing and assessment. Moral issues are presented solely within the contexts of their actual occurrence and therefore require skillful and sensitive identification, probing, and attention to the fullness of each specific context or setting" (246).

This "sensitive identity" and fullness of detail characterize his case. Zaner demonstrates his ability to judge the case because we believe that he has made this "sensitive identification" or in Husserl's terms, the "eidetic vision," the vision of a thing's essence. As with his representation of dialogue there is no separation between Zaner and his characters; they speak together. Look, for example, at Zaner's contention about internal feelings of the son that are simply not a part of the narrative as the case is first presented. "Indeed, the discussions with Jones had first to contend with the heavy guilt the son was already feeling although it was still silence. He was blaming himself for letting 'that damned ER doctor keep her "hooked up"' to the ventilator, a guilt reinforced by Jones's thoughtless comment that 'if you had acted differently, we wouldn't be in this situation now'" (245–46). This statement has multiple quotations within quotations: The narrator indirectly quotes the son who quotes his mother. Yet the larger frame is the narrator's voice, who seems to be able to observe guilt within the silence. Zaner's judgment can only be justified to us by our own faith that he has attained some form of eidetic vision. He demonstrates this with his stylization; he shows to us that he has gotten inside the world of the characters of this dilemma, for his voice has become inseparable from theirs.

Narrative cannot simply present the perspective of another; it can only provide a mimesis of another's speech. We must remember that even reported discourse is mediated representation. Bioethicists have tended, however, to use transformed discourse and only fragments of reported discourse, as seen in the case by Brody and Engelhardt. When reported discourse becomes a featured attribute of the case narrative, it is often a moment of crisis, as if once a patient's voice breaks through the speech of others, it becomes a dramatic rupture in the normal course of medical treatment. Zaner's use of indirect discourse is the exception that proves the rule, for his choice seems to reflect a rhetorical desire to demonstrate the success of his phenomenological method. Unless patients become the authors of their own stories, as has been favored by Arthur Frank, their language will always be mediated through the voices of others. Bioethics have tended to use forms of representation that require that they mediate the perspective of others. One wonders why so few of these ethicists have tried to write not simply about the problems of others but with others.

Chapter Eight: Dax Redacted

"Let it not be said that I have said nothing new. The arrangement of the material is new."
—*Blaise Pascal*

One of the most renowned cases in bioethics is that of Donald "Dax" Cowart, a Texas man who in the early 1970s requested that treatment for burns covering much of his body be discontinued. His case originally came to the attention of many bioethicists through a thirty-minute videotape made in 1974 at the University of Texas Medical Branch at Galveston. This video, titled *Please Let Me Die*, consisted of an interview with Cowart by Robert White, the consulting psychiatrist, and scenes of the treatment of Cowart's burns in a Hubbard immersion tank. In 1975 the case was written up in the *Hastings Center Report* under the title "A Demand To Die" followed by commentaries by White and Engelhardt. After viewing *Please Let Me Die* in a bioethics seminar, Keith Burton, a freelance journalist, met Cowart and proposed making a second video, which was distributed in 1985 under the title *Dax's Case* (Burton 11–12; Kliever xv–xvi). Since the *Hastings*

Center Report's presentation and the two videos, bioethicists have regularly drawn upon Cowart's case to explore the issues of paternalism and competence in medical practice.

This case, I believe, should be of special interest to those concerned with the relationship between narrative and ethics because Cowart's story has not simply been referred to by various ethicists but retold in substantially different ways. The presence of different versions of the same case affords the rare opportunity to examine the relation of re-presentation to representation in a bioethics case. As we shall see, the choices of inclusion and exclusion in the various redactions of this case follow from the needs of the ethicists' philosophical arguments rather than provide a forum for testing those propositions. I am taking—although using in a slightly different manner—the meaning of the word "redaction" from "redaction criticism" within biblical scholarship. Redaction in this critical tradition refers to the way in which the editing of source material can reflect the philosophical or theological perspective of the editor (see Perrin). In order to bring the consequences of redaction into strong relief, I shall examine four retellings of Cowart's story and focus on how the construction of the narrator, plot, character, and point of view within them reveals a rhetorical subtext.

REMOVING THE NARRATOR

In 1982, the narration of Cowart's case in the *Hastings Center Report* was reprinted in a collection edited by Carol Levine and Robert Veatch. In their preface, Levine and Veatch remark that those cases which lack a byline were written by them, the staff of the journal, or by an author who wished to remain anonymous (Levine and Veatch x). In their reprint of Cowart's case with the commentaries by White and Engelhardt, there is no byline. Yet in the original printing of the case in the journal, there is a note set within the text of White's commentary that states that the case "is presented by Robert B. White, M.D., the psychiatrist in the case" (White, "Demand" 10). How, if White is one of the central participants in the narrative and its author, could a reader not immediately discern that he is the narrator of the case? The reason is that the case is written in the third person through an effaced narrator. Gérard Genette distinguishes between two types of narrators: heterodiegetic and homodiegetic (*Narrative* 243–47). Heterodiegetic narrators are those that are not characters in the action described, that is, they are not agents in the story; homodiegetic narrators are characters in the story and can be witnesses to events or the main protagonists. Although the reader would expect that White's case would be told through a homodiegetic narrator, the speaker is a heterodiegetic one.

According to White, the final printed presentation is actually an edited version from a written piece in which he provided a summary of the events as well as his commentary (personal conversation with R. B. White, 7/20/95). The perspective on the case is, nevertheless, quite distinctly that of White. This is most vividly seen when one compares this written version to White's oral description of events in *Dax's Case*. In White's account of events in the video, he emphasizes Cowart's "little boy" anger and the "power struggle" between Cowart and his mother. In the *Hastings Center Report*, all of these characteristics—which in the video are unique to White—are featured as well (see White, "Demand" 9). In *Dax's Case*, however, such descriptions are understood by the viewer to be White's view and not "facts" disengaged from a particular perspective, in White's instance the objective report of a psychiatrist. Interestingly Gérard Genette, in arguing for the importance of distinguishing the author from the narrator in a story, notes that considering them to be one and the same "is perhaps legitimate in the case of a historical narrative or a real autobiography, but not when we are dealing with a narrative of fiction" (*Narrative* 213). This is true in watching *Dax's Case*, where we assume that when each of the participants recounts events, the narrator of their story is the same person as the one speaking, that is, that these are not film actors assuming the roles of the real participants of the case. Yet in the final case presentation in the *Hastings Center Report*, also a form of historical narration and not fiction, there is a dissociation between the author and the narrator due to the creation of a heterodiegetic narrator. How strange it would be, for instance, if in the video White began talking about himself in the third person.

Printed immediately after the case presentation in the *Hastings Center Report*, White's commentary has been kept in the first person. Yet, as I have noted above, the use of a heterodiegetic narrator in the case itself gives an objectivity to White's commentary that is not present in the video. It is only White who sees Cowart's outbursts in terms of parent-child dynamics. Interestingly, in the written case White's comments on Cowart's "childlike rage" and his battles with his mother are presented prior to his (or perhaps one should say the character "White's") entrance into the drama. These statements made by an effaced narrator make his subsequent first-person commentary appear to be the result of knowing the "real story." The consequence of this construction of a heterodiegetic narrator is that White's psychiatric evaluation of events in his commentary is seen as a logical conclusion from the facts. As a result of this editorial decision, White's interpretation of what the narrative is really about—Dax's need for emotional control—makes the most sense to the reader because a heterodiegetic narrator has already presented an interpretation, not as a perspective, but as self-evident data.

THE CASE OF THE MISSING MOTHER

In *Medical Ethics*, Tom Beauchamp and Laurence McCullough present Cowart's case in a chapter on "medical paternalism." They note that they have "developed" the case from its presentation in the *Hastings Center Report*, the first videotape, *Please Let Me Die*, and "an interview with Donald Cowart at the Kennedy Institute of Ethics in April 1983" (80). Here is the beginning of their two-page account of the case:

> In the spring of 1973, a twenty-six-year-old college graduate named Donald Cowart was discharged after three years of military service as a jet pilot. Mr. Cowart had been active in team sports in high school, had performed in rodeos, and was devoted to outdoor activities. He had set his sights on becoming a commercial pilot or lawyer. After his discharge and before moving ahead with his career plans, Mr. Cowart joined his father in a real estate business in East Texas. Two months later he and his father were appraising some rural property about 135 miles east of Dallas. They had unknowingly parked near a leaking propane gas transmission line, and when they returned to start their car, the ignition spark set off a large explosion that engulfed both father and son in flames. (79–80)

On the whole, Beauchamp and McCullough's case presentation follows closely the diction, plot structure, and, frequently, the syntax of White's "A Demand To Die." Like their written source, Beauchamp and McCullough begin with a description of Cowart's life, his love of the outdoors and his recent discharge from the military; they describe the accident and provide a detailed description of Cowart's injuries, his medical care (the Hubbard tankings), his refusal of continued treatment, and finally the request by the medical team for a psychiatric evaluation by White. At times Beauchamp and McCullough are more detailed in their account of events, thickening the plot of the case. Whereas in White's version the reader may conclude that there were only two hospitals involved in Cowart's care, Beauchamp and McCullough relate Cowart's movement from a temporary admission at a local hospital to the Parkland Hospital in Dallas to a rehabilitation center in Houston to a final transfer to the University of Texas Medical Branch at Galveston. Beauchamp and McCullough also add Cowart's attempt to jump out of a hospital room window with the assistance of his room-mate. This narrative incident cannot be found in "A Demand To Die" or *Please Let Me Die* and thus it can be assumed to have come from Beauchamp and McCullough's interview with Cowart at Georgetown University. The authors also continue the narrative beyond the event of White's filmed consultation, for Beauchamp and McCullough conclude their narrative with Cowart's return home.

A closer comparison of the two versions reveals that, although the Beauchamp and McCullough narration seems to follow the plot of White's account, the authors edit out one central character and an entire series of incidents prominently noted in their written source— Cowart's interaction with his mother, Ada. Here, for instance, is Beauchamp and McCullough's handling of Cowart's insistence that treatment be stopped and that he be discharged from the hospital.

> On many occasions he insisted that his treatment be dis-continued. Had he been discharged from the hospital, death by infection was inevitable, but he said he intended to take his life, not to allow death to occur by infection. Mr. Cowart's physical condition prevented his leaving the hos-pital on his own; his discharge *required the cooperation of his physicians.* (81)

The nature of their redaction becomes quite evident when it is compared to the presentation of these events in White's version, "A Demand To Die." In White's narration Ada Cowart is a consequential factor in the decision to be made.

> He became adamant in his insistence that he be allowed to leave the hospital and return home to die—a certain consequence of leaving since only daily tanking could prevent overwhelming infection. The tankings were continued despite his protests. His mother, a thoughtful and courageous woman, was frantic; his surgeons were frustrated and perplexed.
>
> ... He engaged his mother by the hour in arguments regarding his demand to leave the hospital—which, of course, he was physically incapable of doing *unless she agreed to take him home by ambulance.* (9; emphasis added)

Beauchamp and McCullough have rewritten the case in a way that excludes Ada Cowart's role in the narrative. One may argue that it was for brevity's sake that they edited out Cowart's relationship with his mother, but one would then have to respond to why they added a detailed account of his attempted suicide and his roommate's participation. Furthermore it is odd that a narrative that seems to follow so closely the plot of "A Demand To Die" would leave out this figure, who is so central in the earlier version. In the Beauchamp and McCullough narration, it is the physicians who must agree to send him home. In White's version it is Cowart's mother who must agree and cooperate. Beauchamp and McCullough have rewritten the case in such a way that Ada Cowart is not mentioned once. Why would they rewrite this narrative in a manner that excludes Ada Cowart's pivotal role in the plot?

The answer lies in their thesis. As I noted above, Beauchamp and McCullough are using Cowart's case to analyze the issue of paternalism in medicine. The larger thesis of their book concerns the conflicts that can arise between two models of moral responsibility that are each based on different prima facie principles: the beneficence model and the autonomy model (20). In terms of what is the "patient's best interest," the beneficence model is the "perspective of medicine"; the autonomy model is the "perspective of the patient" (22). They argue that "a central task of medical ethics" is "negotiating" between these two principles when they conflict (20). Edmund Pellegrino and David Thomasma note that in Beauchamp and McCullough's text, "The case examples used to illustrate the conceptual content of the text choices are framed almost exclusively as conflicts between beneficence and autonomy" (57). Admitting Ada into the plot of the Cowart case challenges the approach that Beauchamp and McCullough are taking in this text. Cowart's case would have to be seen in relation to another agent outside of the patient-physician dyad. (Interestingly, the final chapter of Beauchamp and McCullough's work concerns "third-party interests," but the two issues they deal with in terms of family relations are "children with life-threatening conditions" and "adolescent sexuality and privacy." In Cowart's case, the health care professionals must respond to a relationship between an adult and his widowed mother; taking the complex issues of such a relationship into account could possibly challenge the principle of autonomy.) White's inclusion of Ada's cooperation emphasizes that the physicians' decisions did not affect just Dax but also his mother. Beauchamp and McCullough redact Cowart's story in a manner that foregrounds the view that conflicts in medical ethics consist solely of the difference between what medicine believes to be the patient's best interest and the patient's desire for autonomy; in doing so they mask Ada Cowart's presence.

A FREUDIAN SLIP

> Mr. L is a twenty-six-year-old single male patient with a past history of intense participation in physical activities and sports, who has suffered severe third-degree burns over two-thirds of his body. Both of his eyes are blinded due to corneal damage. His body is badly disfigured, and he is almost completely unable to move. For the past nine months, he has undergone multiple surgical procedures (skin grafting, removal of his right eyeball, and amputation of the distal parts of the fingers on both hands). He has also required very painful daily bathings and bandage changings in order to prevent skin infections from developing over the burned areas of his body. The future he now looks forward to include months or years of further painful treatment, many additional operations, and an existence as at least moderately crippled and mostly (or totally) blind. (Gert and Culver 152–53)

This is the first paragraph of the presentation of Cowart's case in Charles Culver and Bernard Gert's *Philosophy in Medicine*. The reason Cowart is identified as "Mr. L" is that this was published before he went public with his story; for convenience's sake, I will in my discussion refer to Cowart as Cowart. For these authors, the Cowart case is one of seven cases presented in a somewhat casuistic manner for their discussion of the "justification of paternalistic behavior," and they note that their presentation is an adaptation of White's "A Demand To Die." Culver and Gert retell "A Demand To Die" in a style that seems to be dominated by features appropriated from the clinical case history. This is perhaps most conspicuously seen in that they rewrite White's statement, "He was then twenty-six years old, unmarried, and a college graduate. An athlete in high school, he loved sports and the outdoors" (9), as "Mr. L is a twenty-six-year-old single male patient with a past history of intense participation in physical activities and sports, who has suffered severe third-degree burns over two-thirds of his body." They have also eliminated from "A Demand To Die" how Cowart received these burns. Instead, their plot structure reproduces that of the medical case history (cf. Hunter, *Doctors'*). By identifying Cowart in the first sentence by his age and gender along with a brief summary of his state of being, Culver and Gert are characterizing him through the medical gaze. White's narration of his love of the outdoors becomes translated into a clinical concern with history taking. As in a medical case presentation, there are no agents in the narrative except for White's work as a consultant. Yet what is most striking is how in "A Demand To Die" the block characterization—"a set-piece presentation of a character's traits" (Prince, *Introduction* 10)—is far more complex than that presented by Culver and Gert.

E. M. Forster's distinction of flat versus round characterization has been used often by literary critics to analyze the degree of complexity in narrative portrayals. Flat characters, as the term suggests, are two-dimensional, lacking depth or complexity. Forster states that "In their purest form, they are constructed round a single idea or quality: when there is more than one factor in them, we get the beginning of the curve towards the round" (67). Round characters are complex and multidimensional. Forster gives a litmus test by which the reader can determine if a character is round or flat, and that is if the character is credibly able to produce surprise in the reader (78). White's characterization of Cowart is round, for he is presented in a complex manner. In "A Demand To Die," White includes such block characterizations as: "By his family's account, he was an active, assertive, and determined person, who since childhood had tended to set his own course in life.

What or whom he liked, he stuck to with loyalty and persistence; what or whom he disliked, he opposed with tenacity" and "Although calm and rational most of the time, the patient had frequent periods of child-like rage, fear, and tearfulness" (9). In White's narration Cowart is multidimensional and capable of surprising the reader. In Culver and Gert's narration, however, the reader is given a more two-dimensional characterization. Forster notes that really flat characters "can be expressed in one sentence" (68), and Culver and Gert take all the emotional roundness of White's presentation and in the second (and final) paragraph of their narration give the reader a single sentence: "He has been interviewed by a medical center psychiatrist and found to be bright, articulate, logical, and coherent" (153). I think it would be surprising if this Cowart were able to surprise us. Why have Culver and Gert in their redaction taken the round characterization of their textual source and flattened it out?

To expunge all the psychological nuances seems extraordinarily odd in a text that declares in the preface that the authors wish to show "the value of close and continuing collaboration between philosophers and physicians," especially psychiatrists (vii). White's "A Demand To Die"—the cited source for Culver and Gert's version—explicitly features the psychological aspects of Cowart's character, yet why would they edit out these elements? Culver and Gert argue that the benefit philosophy can have for medicine is that it can provide an "analysis of concepts," and they note "in recent years philosophers have come to realize that medicine, especially psychiatry, employs many interesting concepts that are in need of analysis" (4). Cowart's case is presented as a part of their analysis of the justification of the concept "paternalism." The authors contend that there can be instances of justifiable paternalism when "the evil prevented for S by the moral rule violation be so much greater than the evil, if any, caused to S by it, that it would be irrational for S not to choose having the rule violated with regard to himself" (148). Great stress is placed throughout Culver and Gert's argument on the "rationality" of the person's desires. They argue that in Cowart's case paternalism cannot be justified because Cowart's ranking is as rational as the physicians' (153).

As I have already noted, a significant portion of White's narration of the case is spent showing the complexity of Cowart's state of mind, the roundness of his character. The sentence about Cowart's desire to die, for instance, is given a twist in White's account: He accepted some treatment and then suddenly refused to give permission for the surgery. Recognizing the complexity and thereby the roundness of Cowart's character would in effect challenge Culver and Gert's lucid analysis of the case. In Culver and Gert's account, philosophy can assist psychiatry, but psychiatry seemingly has no effect on philosophy. Philosophy may bring a clarity to psychiatry, but the effect psychiatry would have on philosophy, if they were truly in dialogue, may be to make things unclear as it does in Cowart's case. White's reflections on the case reveals how an understanding of Cowart's psychology, the roundness of his character, challenges our notions of competency. After recalling Dax's "episodic rages," White asks "Was his demand to die a protest against the helpless and infantile state into which he had been forced so many months, a state in which he was totally at the mercy of those who cared for him?" ("Memoir" 16). Culver and Gert's flat characterization is the result of excluding information which could possibly challenge their philosophical assumptions concerning the relation of philosophy to medicine in the resolution of moral problems.

LOSING PERSPECTIVE

> David G's life exploded in flames when he was twenty-
> seven years old. He had come home in May, 1973 from
> three years' service as a jet pilot in the air force to join his
> father's real estate firm. In July the two men went together to
> inspect some property. When their car failed to start, Mr.
> G's father lifted the hood to manipulate the carburetor and
> directed his son to start the ignition. Mr. G did so, and the
> car suddenly was enveloped by fire. (Burt 1)

This is the beginning of the case presentation of
Cowart's story in a chapter titled "David G and Self-
Rule" in Robert Burt's *Taking Care of Strangers*.
Burt's work is an analysis of the way that the law
should function in patient-physician relations. His primary source for his version of Cowart's case is, unlike
that of the previous authors, the videotape *Please Let
Me Die*. In one of his reference notes he mentions that
he is aware of the *Hastings Center Report* narration
(182) and in another that he has been in correspon-
dence with White (122), but he does not seem to draw
upon these sources often in his redaction. (One notable
exception is the use of the same opening image of
Cowart's world "exploding," which was used in "A
Demand To Die.") In an appendix Burt includes the
entire transcript of the Cowart-White interview so that
a reader can verify the manner in which he has represented Cowart's views (2), and he describes as well his
reaction to the visual images of the video. Burt's redaction emphasizes the emotional dimensions of Cowart's
world. He attempts to provide a round characterization. For instance, in response to Cowart's own argument that he be allowed to die because otherwise it
would violate his "civil liberties," Burt counters that we
must place this argument within its "emotional context," for otherwise "we would find ourselves purport-

ing to obey the wishes of a caricature, a cardboard cutout, rather than a fully fleshed and recognizable human being" (6). Burt provides perhaps the most complex retelling of Cowart's case.

After recounting the events of the accident and summarizing the physical condition of Cowart in a non-clinical style, Burt quotes a series of segments from the videotape. The first quotation Burt selects is a lengthy description by Cowart of his hospital entry and his delusional experiences during his first few days; it includes such statements as, "there is no way I could begin to explain the nightmares and the excruciating pain involved in the first events at Parkland that I can barely remember it myself. It was sort of like a dream, and a real bad dream, and I might add that I had a lot of nightmares at the time and I couldn't tell what was really happening and what I was dreaming" (transcript of *Please Let Me Die* [as qtd. Burt 2]). Burt also emphasizes the visual images he sees in the video. He describes how Cowart's "entire skeletal frame was starkly emaciated, looking as if he were a survivor of a Nazi concentration camp" (8). Like his opening image of Cowart's life exploding and the lengthy quotation about his delusional period in the hospital, Burt's association of Cowart's physical state with a concentration camp survivor adds an odd surrealistic dimension to his description.

Unlike the versions discussed above, Burt's narration and his analysis are not separated. In this Burt is unusual, for in most ethics discussions, the case is dissociated from the analysis (as it apparently was done in White's initial telling of the case), but Burt moves back and forth between presenting information and interpreting it. His continual interpretive gloss seems to make Cowart's every statement possess a covert meaning. In the first paragraph, the detailed account of Cowart's actions in relation to his father just before the explosion is later seen as important information for Burt. He places a gloss upon it: "Mr. G's guilt at this fact [that he survived the accident] might be intensified by one particular fact of the accident: he turned the ignition switch that led immediately to the explosion" (10). Following Cowart's statement that he believes ethical lawyers do not wish "to touch" his case, Burt finds a double meaning: "I hear some suggestion in Mr. G's remark of a further question he has about himself, a question of whether 'ethical' people in general 'want to touch' him" (10).

Cowart's story is the first in *Taking Care of Strangers*, which focuses on what Burt believes is the psychological response people have to severe illness. Burt's central thesis in this book is that any laws concerned with patient-physician relationships must take into account that our desire to care for a patient "is intertwined with a wish to hurt that person, to obliterate him from sight" (vi). Burt contends that all the people making these decisions (including the patient, him or herself) have "difficulty fixing" themselves in such categories as "diseased person, mentally normal, mentally ill, retarded, or comatose" and have thereby "tensions between benevolent and abusive impulses" toward themselves and others (vii). In analyzing Cowart's case, Burt argues that Cowart has become "palpably confused about whether he is able to conceive of himself as an individual—as a being separate from others" (11–12). He sees Cowart as an individual who is unable to determine that "he has an identity separate from other people and the external world" (12). Burt's use of surrealistic images and his undermining of Cowart's rational statements through a psychological gloss support his essential belief of the inability of Cowart to distinguish himself from others, to have a realistic portrayal of the world and thus of the consequences of his decisions. In contrast to Culver and Gert's cold, flattened style that promotes the logical qualities of Cowart's character, Burt provides an account in which everything the reader sees is brought into question, every statement Cowart makes has a double meaning. This in turn supports Burt's general thesis of the psychological tensions that exist in such decisions as this.

Burt's account of *Please Let Me Die* appears on the surface to be comprehensive, for along with a complete transcript at the conclusion of the text, he includes vivid descriptions of the videotape presentations of Cowart's tankings. Yet Burt does leave out a series of telling images in this videotape: how Cowart is presented during his interview. The reader would assume from Burt's account that Cowart is shown in a simple full length shot lying in a hospital bed with White by his side, but in fact during the entire interview the camera moves around Cowart's body, creating an odd montage of images. The camera person moves in for tight, close angle views of injured parts of Cowart's body. His hands are given such tight close-ups that at times it is difficult to tell whether they are parts of a human body. The cutting from one part of the body to the next is done so strangely that the viewer has to spend time attempting to figure out what they are looking at. Is it an arm? A leg? Part of his face? Burt provides no description of these images during the interview, images that make Cowart's world seems strange and blurred. It is perhaps the presentation of these images that is the true source of Burt's understanding of Cowart.

All forms of narration are characterized by point of view. Perspective in film and literature, however, should be distinguished, for film uses aural and visual modes of communication in the construction of a point of view (Chatman, *Story* 158). Susanne Langer observes in "A Note on the Film" that for the percipient of a moving picture, "The camera is his eye (as the microphone is his ear—and there is no reason why a mind's eye and a mind's ear must always stay together)" (201). It is just this separation between the eye and the ear that occurs within *Please Let Me Die*. The aural perspective in the video is Cowart's, and, as Burt writes, Cowart's "lucidity, evident in the conversation transcript, powerfully supports Dr. White's conclusion" that he is competent (3). The visual perspective, however, distorts Cowart, showing him through a series of odd angles and warped images. The visual point of view consists of a series of images that contradict and subvert Cowart's rational discussion by providing a (perhaps) unintended commentary on his competency. Burt wishes to argue for understanding Cowart's "real" perspective, an understanding of the true nature of his psyche, but he simply reproduces the point of view of the video. If Burt were to acknowledge the visual perspective, he would in turn have to acknowledge how his understanding of Cowart is mediated through the point of view of this mode of presentation. In the end he excludes from his description the rhetoric of the video, which can persuade a viewer to question the rationality of Cowart's character.

What seems to be a pattern of exclusion in the literary construction of Cowart's story should be particularly disquieting to bioethicists, because I suspect these scholars see themselves as demonstrating the validity of their arguments by applying them to this case rather than demonstrating the limits of their capacity to write objective cases. How is the reader to evaluate a bioethicist's moral paradigm if the ethicist is the one who is judging which particularities are relevant to show? These redactions of the Cowart case reveal that the decisions that determine the inclusion and exclusion of particularities can follow from prior philosophical ideas of what constitutes moral problems and how to resolve them. When one leaves out the mother as a key character, then the reader is unable to challenge one's view of the essential conflict of the case. When one flattens Cowart's character, then one's argument about rationality appears accurate. When one does not include the way one's view of Cowart has been mediated by others, then one does not provide the reader with information that challenges one's psychoanalysis of his needs. I do not wish to suggest that one can include "everything" and become like Borges's "Funes, the Memorious," whose limitless memory was useless because he could not distinguish the meaningful from the trivial. The problem that the various redactions of Cowart's case reveal is not about the issue of selection and exclusion but a far more serious question of the myth that there are clear, unmediated presentations of moral problems. When one presents cases as if they were written by others and are thereby objective data, the reader tends not to question the origin of the data. Bioethicists need to acknowledge that their selection of relevant facts is itself guided by their philosophical perspectives.

Chapter Nine: The Medium Is the Moral Message

Robert Burt's redaction of the Dax Cowart case discussed in the previous chapter raises an additional question of the representation of moral problems. While Burt seems sensitive to the language of the Cowart case, he is unresponsive to the rhetoric of the medium of film.

Most ethics cases do not exist in such a variety of presentational mediums as the Cowart case. As Tom Beauchamp and Laurence McCullough mention, Cowart continues to speak on his case, so it is possible to receive an oral presentation of the case. There are a variety of written presentations of this case, two film presentations, and a laser disc version, which combines written and film presentations. Each of these mediums contributes to how we perceive and understand this moral problem. In this work, I have focused on the literary mode of presentation primarily because I believe this is how the data of bioethics is constructed. I would be amiss not to take into account, however, the way the literary medium conditions the way one experiences case presentations; in this chapter, I wish to briefly consider how oral, written, film, hypertextual, and dramaturgical presentations of ethics cases modify how we judge moral problems.

ORALITY, LITERACY, AND THE QUESTION OF AUTHORSHIP

Bioethics cases are presented principally in spoken or written forms. Since ethics cases in medicine have—as noted throughout this work—an intimate relationship to the world of medical cases, they are a part of the oral presentation of cases within the clinical environment. Ethics cases consequently are presented in grand rounds, during ethics committee meetings, and as part of ethics consultations. A spoken presentation of a case is qualitatively different from a written one. Eric Havelock, in discussing poetry in the ancient Greek world, observes the intimate relationship between sender and receiver in an oral environment, for such a relationship "could be established only by audible and visual presence" (146). In oral communication, no separation exists between the sender and the message, for, without the addition of technology, the message cannot exist outside an intimate, physical contact of at least two individuals. This face-to-face quality of orality, in turn, permits the receiver of the message to invariably also perceive the situated sender. It has often been noted that in medical case presentations the passive voice permits health care professional not to accept responsibility for their actions ("a physical exam was performed"), but this observation sequesters the medical discourse outside the scene of the telling. Hence Renee Anspach, in her analysis of medical discourse, does not seem to see the contradiction of reporting that case presentations are "self-presentations" and then also that the presentations "omit the physician, nurses, or other medical *agents* who perform procedures or make observations." In an oral medium, one cannot separate the message concerning the procedure or observation from the sender as agent, for only in written discourse can the use of the passive voice indicate the refusal to take responsibility for agency. Oral presentations of ethics cases always force the receivers to see themselves in personal relationship to the sender.

Related to this is Walter Ong's contention that one part of the psychodynamics of orality is that knowledge is never disengaged from "human struggle." Ong believes that knowledge within that oral world cannot ever be separated from the ongoing struggle for power within that world, as opposed to the disengaged discourse of writing. For ethics cases this means that in the oral presentation of the message, the receiver must see the message in terms of its relation to the sender and also the receiver's relationship to the sender. Communication that takes place in an oral case presentation exists within the lifeworld of the human encounter. In his analysis of oral storytelling, Richard Bauman notes the layering effect of oral performance. He notes that the sociologist Erving Goffman believed that in every human encounter we are wondering, "What is going on here?" Bauman contends that when we listen to a narrative we are wondering, "What is going on there?" While in a storytelling performance the two questions become entangled, in a case presentation transposed into the medium of the writing, the receiver does not attend to Goffman's question.

In "The Consequences of Literacy," Jack Goody and Ian Watt contend that writing also alters our relationship to the message by contributing a stability not possible with oral forms. Oral societies, because they depend on direct communication between individuals, can lose or transform the past without a sense that anything has been lost. Writing in societies alters this, for these cultures "cannot discard, absorb, or transmute the past in the same way. Instead, their members are faced with permanently recorded versions of the past and its beliefs" (67). It is important to note, however, that post-structuralists like Derrida and Lacan have challenged this belief that the written word is in some manner "stable" and not itself as slippery and complex as its oral cousin. One can argue that prior to the entrance of philosophers into the medical environment, ethics cases were circulated primarily through an oral culture, for only with the public writing of cases (which in turn become part of a consultable past) can bioethics become a field of academic study. When ethics cases remain part of an oral tradition, they are subject to the transformation and forgetting that occurs in that medium, and when they enter into the medium of writing are in turn subject to scrutiny and challenge.

THE FILMED CASE

In the so-called "Johns Hopkins case," the parents of a child born with Down's syndrome denied the child relatively routine surgery. The child died in the hospital because the health care professionals honored the parent's wishes. A film dramatization of this incident was made, titled "Who Shall Live?," and it seems that some who watched the film believed they were watching a documentary of the events as they unfolded. David Rothman, for instance, observes that the "press reports on the film either missed the fact that this was a re-creation or took the baby to be the actual baby" (R285). Part of the power of film is that it grants the illusion that we have become witnesses to the events portrayed, and we tend to watch filmed events with less skepticism about their constructed nature. For example, with the first film made of the Cowart case, I argued that Robert Burt seemed unaware of the degree to which he was affected by the way the film was shot.

Yet a brief analysis of the second Cowart film, *Dax's Case*, confirms that the medium of film must be viewed as critically as we do other mediums.

In *Dax's Case* the segments of interviews are often juxtaposed in a way that makes it appear as if the participants are in dialogue with each other. There is a great irony in this, for at no point in this video do the participants in the decision concerning Cowart's care talk to each other. For instance, when the surgeon in the case mentions confronting Cowart, he recalls saying to him, "Don't ask us to let you die, for in a sense that means that we are killing you." After telling Cowart that he can kill himself when he has some use of his hands, he reiterates, "but don't ask us to stand here and ask us to literally kill you. You want to die, you do it yourself." Then the next shot is of Cowart calmly stating: "The argument that not treating a patient is the same as killing borders on the ridiculous. If letting the patient die is characterized as playing God then treating the patient to save his life has to be as well." The editing throughout *Dax's Case* paces the interviews as if the various participants are responding to each other's arguments, and usually Cowart is given the final word. The sequence of editing and the final montage of the video present an argument and a perspective apart from that of the participants.

The academic study of film has often drawn upon the tools of semiotics, for the idea of systems of signs is useful in understanding a medium that is able to rigorously control the channels of sight and sound (see Stam, Burgoyne and Flitterman-Lewis). In order to interpret the films of the Cowart case one must take into account the various ways sight and sound are used to create a particular "take" on events. Sound is a fascinating element of film, for it can encompass a mix of voice-over narrators, dialogue, "accidental" noise in the scene, and inserted music. Howard Brody provides a sensitive reading of the various sounds in the first film of the Cowart case: "A radio blares in the background, machinery whirs, and the place seems more like an automobile repair shop than a sanctuary for suffering humanity" (*Healer's* 75). Just as we can distinguish narrators by their relationship to the diegesis (whether they are a part of or apart from the story world), David Bordwell and Kristin Thompson suggest that sound within a film can be:

> *simple diegetic sound*: sound that has its origin from within and temporally with the story,

> *external diegetic sound*: sound from within the story world which the characters are aware of,

> *internal diegetic sound*: sound within the mind of a character,

> *displaced diegetic sound*: sound from within the story world but not temporally with it,

> *non-diegetic sound*: sound outside the world of the characters. (as cited in Stam, Burgoyne and Flitterman-Lewis 60)

In the first Cowart film, Cowart's voice sometimes occurs with scenes of his tankings. The health care professionals taking care of the tankings do not have access to this sound, so it is an example of a displaced diegetic sound. The use of displaced diegetic sound permits Cowart to make a commentary—a remarkably rational and thought-out one—that exists in contrast to the later sounds (simple diegetic and external diegetic) of the apparently insensitive workers and his own screams of pain. Sound never has a "natural" form within film and like the other semiotic features of a film (lighting, framing, camera angles, editing, color), affects how we understand the narrative presented.

TOWARD THE HYPERCASE

The Center for Design of Educational Computing at Carnegie Mellon University have converted the Cowart's case into a new medium, which combines many of the elements of the oral, visual, and written forms in which it has previously been shown. This videodisc, *A Right To Die?*, not only provides a multimedia presentation of the events, but more importantly for our concerns in this chapter, it also offers the capacity to view a "hypermedia" version of the case. The term "hypertext," coined by Theodor Nelson, refers not merely to a computer's capacity to provide an interchange of different mediums but an entirely different way of reading a narrative. Hypertexts are instances of "nonsequential" texts, and what is referred to as a hypermedia system involves the extension of this lack of sequentiality to nonprint mediums, such as video and sound. Unlike reading a printed book in which the text must be physically arranged, a computer permits a writer to construct a story so that the reader is the one who decides the ordering of the events. When one reads or views a hypertext or a hypermedia system, there are no predetermined ways of moving through the narrative; instead, there are a multiplicity of possible entrance points and routes through the story. Subsequently, reading a hypertext, such as Michael Joyce's *Afternoon, a story*, provides an unusual experience, for the remarkable freedom (and at times lack of guidance) that one has interacting with the text. For this reason some critics assert that the traditional notions of plot have begun to be challenged by hypertexts and hypermedia systems (see Barrett; Delany and Landow; Landow, *Hypertext*). A hypertext or hypermedia system essentially furnishes the reader a story without a predetermined plot, for each encounter with the text can result in a different arrangement of the events by the reader.

The creators of A *Right To Die?* give the reader two separate "pathways" in which to explore the Cowart case. The first pathway is through a section called "Guided Inquiries" and the second is through the "Archives." The Guided Inquiries pathway entails a series of questions to the receiver and clips from the film. The Archives pathway comprises just the clips without any determined order of viewing for the receiver. These pathways should be distinguished from each other primarily through the different ways in which they present the Cowart story through the technology of the computer. The Guided Inquiries pathway, I contend, should be understood as a multimedia presentation and the Archives pathway as a hypermedia one. The creators contend, "The basic purpose of the videodisc is simple: to supplement traditional abstract case narrative or linear video media for presenting case studies" (8). Thus they are aware that the uniqueness of a computer presentation of the Cowart case lies not simply in being able to move easily from texts to visual material but in its capacity to present the case in a nonlinear manner.

My view, however, is slightly different from that of the videodisc creators, who think that "interactive" is the opposite of "linear." I think the contrasts should be understood as "interactive" versus "passive" and "linear" versus "nonlinear." This mistaken belief on their part leads them to argue that every experience with the videodisc is nonlinear, yet it is only in the Archives pathway that the reader is challenged with a nonsequential narrative.

The Archives pathway gives the reader little guidance in respect to how to examine the material. As the written guide explains:

> The archives section of the program allows you to move independently through the case narrative and all audiovisu-

al material. In the archives, you may review the case summary without being prompted for a position, and the audiovisual materials are organized around the issues and principal characters of the case. The purpose of the archives is to provide you with the means to carry out self-direct exploration of the facts and the issues of the case. (15)

The main menu of the Archives section consists of 1) the case, 2) the principals, and 3) the issues. The introductory summary of the case is quite brief.

In 1973 an unmarried 25-year-old man called Dax returned from active duty in the Air Force and while waiting for the airlines to begin hiring, worked in real estate development. In July of that year, Dax drove out to look over a bit of unpopulated land that he was considering as an investment. As he was getting ready to leave, his attempt to start his car set off an explosion in a buried but leaking gas pipeline. Dax received second and third degree burns over two-thirds of his body.

Following this written presentation the reader is left on his or her own. Unlike the Guided Inquiries pathway option and all other traditional case presentations, readers do not know where the case presentation begins or ends. The reader's role in a hypertext or a hypermedia system is substantially different from that in a traditional text. The blocks of material that make up a hypertext or a hypermedia system are commonly referred to as "nodes," and they can comprise any audiovisual material that can be digitized. Nodes are linked together to allow a reader to move from one to the next without reference to a sequential ordering. John Slatin notes that, "A node cannot, by definition, be entirely free of links—a node is a knot, is always embedded in a system—and that connectedness in turn gives the node its definition" (162). In the Archives pathway, the nodes consist of the blocks of video material that have been culled from the two videos of the case. Scholars of hypertexts or hypermedia systems sometimes refer to the reading experience as "exploring" or "navigating" through the text rather than as "reading." Yet this metaphor does not, I suspect, accurately reflect the reader's task, which is not merely to move from node to node but is also to remember and to connect nodes together. This is, of course, true of all reading experiences, a movement of recall and anticipation, but in a nonsequential text, the reader is the one selecting what will be recalled and to what other texts it will then be connected. By moving through the nodes, a reader is creating a specific sequence, which may be different from that of other readers or from the way that reader would concretize the text in a second or third navigation. The difference between this and traditional narratives is that in the hypertext or the hypermedia system the sequence is determined not so much by the author as it is by the reader. Selecting from the nodes gives the reader the role of the editor, that is, the reader decides how the segments will be put together. In discussions of hypertexts this is sometimes referred to as collaborative writing (Landow, *Hypertext* 88–94), but for a hypermedia document it is perhaps more accurate to refer to it as an instance of collaborative editing. The editor of a document must rely on the material that has been furnished by others and all material is presented with some slant; there are no unbiased facts. For example, the nodes of Cowart's perspective are all derived from the first video, *Please Let Me Die*, and none of the interviews with Cowart ten years later from *Dax's Case* have been included; the creators deprive the reader of a large number of the video segments of Cowart's later perspective on events. One suspects that the including of the videotape of the posttreatment Cowart might have influenced the reader who is not aware of how this case turned out. If the reader knew, for instance, that Cowart had survived and was able to participate in an interview then the reader might be swayed from agreeing that treatment be discontinued.

I do not wish to suggest that hypertexts or hypermedia systems lack a semblance of structure or leave a reader without any direction, for there are currently textbooks designed to assist those writing hypertexts or hypermedia systems in creating a particular reading experience (Berk and Devlin). The nodes can be organized along various lines. In *A Right To Die?*, the reader can either pursue the case through the "Principals" or through the "Issues." In the Principals section there are series of nodes for each of the principal participants within this case including Rex Houston (his lawyer), Ada Cowart (Dax's mother), and Leslie Kerr (a nurse). Just as one selects one of the principals, one can choose which nodes to hear. The reader could, for instance, easily listen to only the health care professionals and ignore Cowart's perspective.

This freedom given to the reader profoundly affects some of the narrative elements of a case presentation already discussed in earlier chapters, such as narrative discourse and closure. In his analysis of the structure of narrative discourse, Gérard Genette noted the importance of what he referred to as "frequency" (*Narrative*). Although in the story something may happen only once, in the discourse of the narrative the event may be retold several times, and the frequency can affect the reader's experience of the narrative. Since the narrative discourse in a hypertext is an act of collaboration between writer and reader, frequency takes on a different meaning (Liestol 96). Frequency has been an aspect of ethics case presentations as well. For instance, in the *Hastings Center Report*'s written presentation of the Cowart case the events of the accident are told twice. First in the opening line: "Two months after being discharged from three years of military service as a jet pilot, the world of Donald C.

exploded in a flash of burning gas." Then further along in the presentation is the sentence: "Later when they [Cowart and his father] started their automobile, the ignition of the motor set off a severe and unexpected explosion." In the hypercase, frequency is under the control of the reader, who may wish to hear the same account or point of view several times. Furthermore, the reader can decide which aspects of the story are not worth listening to.

Closure is another key feature in understanding the structure of a plot and, as with frequency, a hypertext or a hypermedia system can alter aspects of narrative (Douglas 159–88). Barbara Herrnstein Smith's characterization of this phenomenon as the reader's "expectation of nothing" takes on new meaning. In the Archives pathway, however, the reader is not prompted to select or ignore nodes nor ever told when to stop. Does a reader have to watch all of the nodes or just a selected number? How does the reader decide when they have reached stasis when the narrative literally does not end? As J. Yellowlees Douglas has noted, "Where print readers encounter texts already supplied with closure, an ending, readers of interactive fiction generally must supply their own sense of an ending" (164). With no preset closure within the case that informs a reader that they need not explore the case anymore, how does a reader decide how many principals in the case they need to hear from in order to fulfill the task that a decision on the case be reached? The reader could essentially create different readings of this case and as many different ones as they wish. The reader's own interests will affect not only to whom they listen but how many times the reader attends to their message. What if the reader only has a limited period of time to decide what to do in the case? All of this is left up to the reader, and the reader is responsible for deciding when enough information has been received.

COOLING THE CASE

Hypertext raises important issues concerning the degree of activity demanded of the receiver in various mediums. Marshall McLuhan's well-known distinction between hot and cool media relies primarily on the degree of participation a receiver needs in order to make a medium understandable. McLuhan argues that hot media (such as film) require less participation by the audience to "complete" the communication than cool media (such as television). The hypertext form of the Cowart case would be considered within this scheme as a very cool medium, as opposed to the hotter medium of the video versions that require a lesser degree of participation. Yet one could argue that as cool as the hypertext is, there is a way to bring the temperature of the case down further if desired. This, of course, assumes that cool media are more valued than hot ones since, I suspect, we generally presume that a medium that requires more participation and activity by the receiver is far more desirable than passive ones because they permit a questioning of the reliability and raise awareness of the constructed nature of cases.

This super-cool medium is performance. A novel way of conceiving of ethics cases is to view them as scripts that will be performed. Conceiving of a text as "script" rather than a "work," means that the audience cannot receive it until they enact it. I submit that if authors knew that their cases would have to be performed by the receiver, it would compel the sender to enter into a highly critical mode about the reliability of the case. I take this idea of using performance as a way of critiquing texts from the work of Victor and Edith Turner. The Turners argue that performance can suitably critique ethnographies by revealing how they "depart from the logic of the dramatic action and interaction they have themselves purported to describe"

(140). In a similar way, ethics cases written in anticipation of performance by others require descriptions of
moral problems that give future directors an ability to
envision the flow of the drama and future actors insight
into the motivation of the characters. Furthermore, the
sender, by presenting cases in this form, anticipates
critiques of their reliability and authenticity by the
receiver.

Performance also permits the receiver to engage in a
medium that includes a multitude of different senses
previously not a part of the traditional case presentation. While multimedia presentations permit an interaction of visual images, texts, and sound, performance
entails these senses along with smell, taste, and touch.
Performance permits a far richer case than just text
itself. This in turn allows us to provide "an alternative
to the atemporal, decontextualized, flattening approach
of text-positivism" (Conquergood 189) that has been a
part of the presentation of data in the Western acade-
my. It seems odd that a discipline like bioethics that
has striven to relate abstract theory to the body has
remained uncritical of the disembodied way cases are
presented. Performance can become a way not merely
of making a case more interesting but a way of knowing distinct from the more passive textuality that has
dominated bioethics.

Each of the mediums discussed in this chapter,
thus, conveys an epistemology, a way of knowing the
moral world. The difference between orality and literacy creates questions of authorship that we tend to forget in the way bioethics has been print-dominated. The
filmed case, although it appears to present the world in
an unmediated manner, is actually dependent on a web
of codes that do not so much make the world more
available as simply add to the rhetorical complexity of
moral messages. Hypertext and the case-as-script are
mediums that demand a level of participation by the

receiver of the message previously not asked in the ways cases have been presented. Both of these forms could be considered better than the traditional forms if we thought activity was a desired goal of moral investigation. Although I do not wish to be viewed as advocating technological determinism—the notion that the medium of presentation necessarily shifts our way of thinking—the discipline of bioethics must, I believe, take into account how the rhetoric of the medium influences our perspective on moral events.

Chapter Ten: Sexing the Case

"It is impossible to *define* a feminine practice of writing, and this is an impossibility which will remain, for this practice can never be theorized, enclosed, coded—which doesn't mean that it doesn't exist."

—*Hélène Cixous*

A remarkable characteristic of medical humanities, but one not often remarked on, is the way the two dominant disciplines in the field—ethics and literature—are often divided by gender. I do not mean to suggest that one does not find women "doing" bioethics or men working in literature-and-medicine, but rather that each discipline is predominately of one gender. The key figures in bioethics have tended to be men; those in literature-and-medicine have tended to be women. This difference, I believe, has its roots in larger cultural issues.

Arthur Kleinman has commented on the differences in the "cultural logic" in divisions within North American medicine of "hard" and "soft" (30). He, a psychiatrist himself, observes that certain specialties in medicine are deemed "harder" (surgical subspecialties, pathology) than others (psychiatry, family medicine). Specialties that involve "talk and cognitive activities" are considered "softer" than those specialties that focus on "procedures that enter the body." Medical humanities, in general, would be placed on the soft side of this division, softer than psychiatry; within the field, ethics would be considered "harder" than literature-and-medicine. Thus it is easier to get funding to start an ethics center than a literature-and-medicine center, and I suspect that literature-and-medicine scholars tend to be the last hired and first fired within a medical school. Kleinman notes that in medicine the soft specialties tend to attract women and the hard men, and this division has a monetary corollary as well. The gender division in medical humanities may be simply mirroring the general cultural logic of its culture of origin.

I wish to show in this chapter that this gender division in medical humanities has had a profound consequence for the way we analyze ethical problems. There has been much work in feminist epistemology and its relation to moral decision-making. Most renowned has been Carol Gilligan's discovery of the "voice of care," which she believes expresses an undervalued way of moral knowing in the justice-fixated tradition within studies of moral development. Bioethicists have developed this insight into a critical tool in the examination of moral decision-making and the framing of issues within the field (see Crysdale; James Lindemann Nelson; Sherwin, "Feminist"; Sherwin, Longer). Although I find this criticism of moral epistemology persuasive, I am interested in how a gendered way of knowing has influenced the ethics case. If cases are the data of bioethics then we need to understand how the tendency of this data to be constructed by men has affected the discipline. In this chapter, I apply to ethics cases some of the insights into narrative theory that have been raised by feminist critics. Cases, I submit, are written in a way that has unintentionally privileged a male worldview.

THE GENDER OF THE VOICE

Some narrative theorists have contended that the issue of the gender of a text's voice is as important a distinction to make as whether the speaker is homodiegetic or heterodiegetic, reliable or unreliable, witness or protagonist. Lanser argues that attention to the issue of gender in a narrative is necessary for the impact gender has on the response to the text. When the text does not explicitly inform the reader of the gender of the speaker, then Lanser argues we assume that the speaker is male.

> Sex differences ... permeate the uses of language and condition the reception of discourse; along with other social identifiers marking the relationship of a textual personage to the dominant social class, sex is important to the encoding and decoding of narrative voice. Because while heterosexual males of a certain socioeconomic position constitute the dominant class in Western society, the unmarked case for both writing and narration is the male case: writers and narrators are presumed male unless the text offers a marking contrary, as they are also presumed white, heterosexual, and (depending on the period) upper or middle class. (166–67)

Lanser's assertion has not been received uncritically by her field of narratology. One of the goals of narratology has been to construct a "science" of narrative, and this telos has resulted in some critical responses to those like Lanser who wish to raise the question of context (see Chatman, "What"; Prince, "Narratology"). This is primarily because such an analysis appears to introduce interpretation, and thus the dreaded subjectivity, into the field. Lanser and others have challenged the sharp divide between the science of narrative and the science of interpretation. When we read narratives we bring with us assumptions concerning the speakers of texts that go unmarked; therefore to deny the contextual features of narrative is to ignore some fundamental structural elements simply because of a limited (and perhaps unself-reflexive) understanding of science and of the possibility of objectivity.

In a similar way I wish to suggest that the reading of an unmarked ethics case evokes the male voice. There are two reasons this should be. First, ethics cases tend to adopt the style of the medical case presentation. This results in their being read much like medical cases and the implied voice of medical presentation carries over into the ethics case. The medical case is itself conventionally unmarked, but the traditional division of labor in Western health care—physicians: male and nurses: female—means that the voice of the medical case echoes with the established male domination of the profession. Second, the predominance of men in bioethics reinforces the cultural assumption that a voice that lacks marking is male. This could be contrasted with those genres in which we assume the unmarked voice to be female, like romance novels and women's magazines.

Look, for instance, at a case from David Thomasma and Patricia Marshall's textbook *Clinical Medical Ethics*:

> Mrs. Egan is a 36-year-old mother of four who, during her last delivery, suffered a tear in the opening of the cervix. At that time, the physicians told her, should she want to have another child, she would need to have a "purse string" or cerclage operation, by which the cervix is sutured to give it sufficient structural strength so as to avoid a spontaneous abortion. Mrs. Egan declined, saying that she wished to have no further children and therefore did not need to have the operation.
>
> Two years later the woman's physician finds that Mrs. Egan is again pregnant. The physician recommends that the woman have the cerclage operation, and the husband, Mr. Egan, concurs, insisting that he does not want the fetus

aborted. The husband hopes it will be a son, for whom he has long waited. The dispute continues into the fourth month of the pregnancy, at which point a sonogram shows that the fetus is a boy.

Mrs. Egan still refuses the cerclage operation, arguing that it is her right to have an abortion until the fetus is viable, and it should not concern anyone how she goes about getting it done, either by medical intervention or because of the structural weakness of her cervix. (24)

At the end of the questions appended to this case, the authors state that this case was "[u]sed with permission of Charles Webber, Ph.D.," yet since Webber is not the physician in the case and no mention is made of an "ethicist" involved in the case, the reader is unable to determine his relation to the events narrated. The reader does not know who is the biological author of this case, even though the biological authors of the textbook are a man and a woman. Following Lanser's argument, a text that does not mark the gender of the narrator, readers in this culture will assume the narrator is male and, as I have noted, this is especially true of the genre of the ethics case. The relationship, I contend, between the gender of the narrative voice and the gender relations in this case are of central importance in our understanding of the events. If the physician's authority has a relationship to the male authority within our culture, then the kind of authority that Mrs. Egan asserts in this situation becomes entangled within the implied gender of the narrator. Especially because this case concerns issues of a man wishing to have decision-making power over a woman concerning the birth of a son, a male narrator's voice influences the reader's perspective on the moral issues.

EN-GENDERING THE PLOT

Although it is difficult to situate the narrator of Mrs.

Egan's case, one can easily identify it as an ethics case. It possesses a plot that readers expect in ethics cases, a dramatic moment that breaks the ordinary patient-physician relationship—here, the drama of the patient's husband's appeal. The work of Claude Bremond best illustrates the analysis of narrative as a series of dramatic possibilities, a series of sequences in which a choice has to be made between different alternatives. In medical ethics cases, the point of choice is always a moral one. Bremond's analysis entails showing the series of possibilities, the selection of one choice, and the final success or failure of that selection.

But this definition of plot through choices has been challenged by some feminist theorists as an expression of a male way of thinking. For example, Anne Cranny-Francis refers to "quest narratives" as a male concept that should be seen as opposed to concepts of narrative that are nonlinear and cyclical. Josephine Donovan, writing on the work of Sarah Orne Jewitt, argues that Jewwitt's writing has a "plotless" structure, which "is an essentially feminine literary mode expressing a contextual, inductive sensitivity, one that 'gives in' to the events in question, rather than imposing upon them an artificial, prefabricated 'plot'" (271). The idea of narratives that lack the imposition of an artificial, linear plot upon human experience in turn challenges some of the most fundamentally held ideas in narrative studies. Yet if one creates a model of narrative based upon male stories, one might create circumscribed models that then define what is and is not a narrative. The inductive model created to describe all narrative becomes in turn a deductive device that prescribes what is and is not considered the data for analysis.

Does bioethics provide us with ethics cases that plot in a feminine style? Rita Charon's case narrative "A Relative Stranger" provides, I believe, just such an example. The physician-narrator of Charon's case recalls the care of Mrs. Kahn, a woman in her eighties with a variety of health problems. Mrs. Kahn brings with her an entire array of physical and social problems that the physician must respond to.

> I found it difficult to take care of her. A routine visit took twice the time of other routine visits because of the communication problems and the challenge of sorting through her medical regimen every time she came. I ended up doing her pelvic examinations and flu vaccinations because of her unwillingness to see other practitioners. Her refusal to be evaluated for the hearing and dental problems annoyed me, if only because fixing those problems would have made my task easier. I felt "stuck" with a patient whose health status would predictably deteriorate in the foreseeable future and whose care would involve me more and more in time-consuming and unrewarding management. I could already predict the difficulty of placing this patient in a nursing home or trying to arrange home-caring nursing for her. As I invented an unpleasant future for myself and Mrs. Kahn, my resentment over having been "dumped on" increased. (52)

Her case seems closer to the fiction of Sarah Orne Jewitt than to what we normally think of as an "ethics case." In many ways, some might not perceive Charon's narrative as an "ethics case," for there is no moment of decision in a Bremondian fashion. To say that "nothing happens," however, seems an odd way of characterizing the story. In some ways, the problems that Charon's narrative explores are more fundamental to the daily practice of health care professionals than the way ethics cases tend to portray these issues. Drawing upon the work of Kathryn Allen Rabuzzi, Donovan observes that in narrative representations women's experience "appears ... static, and in a mode of waiting. It is not progressive, or oriented toward events happening sequentially or climactically, as in the traditional masculine story plot" (219). Charon's narrative provides a rich examination of the moral problems of the everyday and, as Donovan suggests of Jewitt's plots, it provides an exploration of the cyclical nature of the moral work of medicine.

In contrast, the tendency of bioethics to be focused with plots of climax may represent a male epistemology within the field.

CHALLENGING THE BINARY CODE OF BIOETHICS

In their general analysis of ethics in medical practice for a book on ethics in obstetrics and gynecology, Laurence McCullough and Frank Chervenak identify the areas of ethical conflict as including "[i]mplementation of management strategies based on clear-cut beneficence-based clinical judgment ... vs. implementation of management strategies based on deliberative interests of the patient" and "[a]utonomy-based management plans vs. interests of third-parties" (72). Their text exemplifies the tendency to frame ethical issues through binary splits of clear-cut opposing principles or values. They are not unique: Many guides for resolving moral issues begin with identifying the principles in conflict. Although bioethics in the United States has been framed by the four principles of beneficence, autonomy, justice, and nonmaleficence, in a moral analysis one is expected to select two principles that are in conflict and argue for the dominance of one of them. The assumption is that there would not be anything to discuss in bioethics if there were not a clash between two of these moral duties.

Hélène Cixous has argued that the tendency toward binary construction is a result of patriarchal thought systems. In opposition to those structuralists who argue that binariness is necessary to produce meaning, Cixous challenges the necessity of opposition for definition. Cixous provides a series of examples of binary oppositions:

Sexing the Case

Where is she?

Activity/Passivity

Sun/Moon

Culture/Nature

Day/Night

Father/Mother

Head/Emotions

Intelligible/Palpable

Logos/Pathos (Cixous and Clement 63)

Cixous argues that the "master" binary pairing governing these binary splits is male/female and that throughout this tension, the male is always seen as superior to the female. Yet she imagines the possibility of seeing a multitude of signifiers that do not exist in simple opposition to one another. These binary pairings raise two distinct issues. First, in the opposition women are placed in the subordinate role, and the outgrowth of this hidden hierarchy is that all things on the right side of the equation become devalued. Second, this tendency toward creating oppositions is considered by some feminists as the expression of a male worldview. Cixous argues for *l'écriture féminine*, feminine writing, which would playfully resist simple binary oppositions and stasis in discourse.

In bioethics the standard opposition presented is autonomy/beneficence, and, although some wish to counter the tendency, there has been a general consensus that in this conflict autonomy takes precedence. I suggest that our tendency to create case narratives out of this opposition is a reflection of the binary split of male/female. Autonomy (self-rule, independence, rights) is a masculine trait that is seen to be dominant over the feminine trait of doing what is best for the other, something that has been challenged by those in care ethics. Are there examples of a "feminine case," that is, a case that could be designated as *l'écriture féminine*? At the time of this writing, I am not aware of such a case, but I believe to pursue writing a case that, like Cixous's own writing, which could blur the lines between analysis and poetry, ethics and aesthetics, paradigms and narrative is a worthwhile endeavor.

Private Cases

Lanser has also argued that a feminist narratology should include in its understanding of the narrative person (i.e., first-person vs. third-person narrators) a division of public versus private texts: "By public narration I mean simply narration (implicitly or explicitly) addressed to a narratee who is external (that is, heterodiegetic) to the textual world and who can be equated with a public readership; private narration, in contrast, is addressed to an explicitly designated narratee who exists only within the textual world" (461). Lanser observes that this is simply adding to the typology of narrative theory, a distinction that has long been within feminist literary criticism, for there has been a tradition of awareness that many of women's texts were written only for a private audience. Lanser wishes to acknowledge that the difference between private and public narrative levels needs to be taken into account when analyzing narratives.

This distinction between the private and public needs to be applied to ethics cases as well in order to understand that certain cases never reach public attention because the teller does not have ability to bring them into the public sphere. An ethics case narrative by Carole Warshaw and Suzanne Poirier illustrates this well. The case titled "Hidden Stories of Women" recounts Calvin Flowers, a medical intern, attempting to understand the reason for the high blood pressure of Vera Miller, who comes to his clinic. After not being able to determine the physiological or social cause of her illness, Flowers and Miller playfully wonder what type of things would be revealed if she were placed under hypnosis. Slowly Miller reveals a long history of physical abuse:

> "What started first? He started staying out late?"
>
> "He started staying out," she confirmed, "then all the other little things started. He was working but he don't want to bring home any money, left me to try to get on aid. Things started going downhill." After she tells Dr. Flowers they had been together for 25 years, he asks, still unable to register fully what he has heard, "All during that time he had never hit you or anything like that?"
>
> This time, Mrs. Miller doesn't hold back. "I remember so many beatings I wouldn't know where to start." (52)

The case provides an example of a physician stumbling upon a private case, one that tends not to be found in the bioethics literature. It is a case where traditional bioethics provides no simple solution; it does not involve a clear conflict of values but a need for a moral hermeneutics that can respond to the issue of abuse in male-female relationships. As the authors of the case observe, there is a hidden case within the medical case of Miller's uncontrolled hypertension. In some way the hypertension is a public case while the story of her abuse is a private case, one that traditional medicine and medical ethics do not often recognize.

I have in this chapter suggested some ways in which the data of bioethics has been influenced by gender. Narrative is a cultural form that carries with it the implicit cultural codes of its origin. Those elements of narrative that many have come to consider "natural"— or at least natural for the European worldview—may be subtly influenced by gender. It is only in critically examining voice, plot, binary opposition, and privacy that we can gain insight into how gender divisions affect the data of the discipline. I would like to conclude that this exercise should be done with other elements of critical theory: We need to race the case, queer the case, and class the case. In a field that has been so openly dedicated to the issues of social justice, understanding how social injustice covertly enters the field's data seems a necessary critical move.

Concluding Remarks: Taking Stories Seriously

"Every style is a means of insisting on something."
—*Susan Sontag*

In the last chapter I mentioned the hard/soft distinction that exists in relation to bioethics and medical humanities. Consequently the entire idea of "narrative ethics" has struck some as an attempt to sneak the soft field of literature-and-medicine into the more rigorous field of bioethics, for it confuses the subjective world of the arts with the objective goal of philosophy (see Clouser, "Philosophy"; Hawkins, "Literature"). I wish to challenge this assumption by showing that the concerns of literature-and-medicine, especially with narrative, are not peripheral or supplementary but at the nerve center of the field. My interest in analyzing the narrative features of bioethics cases has not been to reveal quirky or simply interesting elements of the field. Revealing the parallels between a Robert Veatch case presentation and a Raymond Chandler novel has been a serious concern with the entities considered the data of this academic field.

One response to this argument (assuming that it has been persuasive) is that bioethics might as well abandon the case: Why bother anymore with cases if they simply are in the end rhetorical fodder for the philosophical frames of the teller? I think such a conclusion would be a mistake, something akin to throwing out the philosophical baby with the narrative bath water. The dichotomy that I have been interested in attacking has not been that there is no real difference between a real case and a fictional one—a difference between unconstructed and constructed cases—but that this difference in the end does not matter. Real people and real events must be turned into representations when they are transformed into narratives. I believe, however, that bioethics enters into an important area when it attempts to apply the ideas of moral philosophy to plausible narratives of moral problems, especially in the bioethics movement's engagement with *phronesis*.

Although the concept *phronesis* has its roots in Aristotelian philosophy, it has come to be used in bioethics as the application of moral philosophy to specific circumstances. The idea has been drawn upon by proponents of quite different approaches in bioethics; phronesis has been especially prominent in the work of scholars who have recently criticized principle-based ethics. John Arras, in promoting "the new casuistry," asserts that it "can be viewed as a defense of the Aristotelian virtue of phronesis (or sound, practical judgment)" (38). In arguing for ethics centered on virtue, Edmund Pellegrino and David Thomasma refer to phronesis as "medicine's indispensable virtue" (84–91). Alisa Carse, advocating a care-based approach, criticizes an abstract appeal to principles that ignores the central place that "a discernment of the particulars" must have in bioethics (12). And Kathryn Montgomery Hunter, in writing about narrative ethics, sees an essential connection between narrative epistemology and phronesis ("Narrative"). In response to some of the criticisms of principle-based ethics, however, Tom Beauchamp and James Childress in the fourth edition of *Principles of Biomedical Ethics* observe that, "We come to understand principles, and what they exclude and include, by making judgments in particular circumstances" (4th ed. 107).

The assumption of many principle-based advocates and their critics is that ethicists demonstrate the validity of their philosophical approaches through such practical judgment, and the reader in turn is supposed to be able to judge whether they have been successful. Yet how is a reader to evaluate the ethicist's phronesis if the ethicist is the one judging which particularities are relevant to the case? The decisions that determine the inclusion and exclusion of particularities can follow from prior philosophical ideas of what constitutes moral problems and how to resolve them. In such instances, the case does not challenge the philosophy but instead rhetorically supports the abstract arguments of the ethicists. David Burrell and Stanley Hauerwas observed in their classic article, "From System to Story," that our moral notions, "do not merely describe our activity; they also form it" (119–20), but ethicists have tended to maintain the fact/value distinction in their presentation of cases. This is perhaps most vividly illustrated in the construction of a case narrator from the philosophy narrator, with styles so different that one automatically assumes that the two are distant relatives rather than conjoined twins. As Wayne Booth notes in *The Rhetoric of Fiction*, narratives contain rhetorical strategies by which the teller "tries, consciously or unconsciously, to impose" a world upon the reader (xiii). I believe that the way out of this problem is not to abandon case narrative but instead to be aware that when ethicists write cases, they are rhetorically imposing a world upon us, a world that excludes as well as includes those particularities that allow us to make the best possible moral decisions. What is at stake here is our ability as readers to verify the arguments of bioethics in the way that its practitioners have claimed it can be verified. This raises troubling questions about what we mean when we refer to bioethics as an "applied" discipline. If bioethics is to continue to rely upon testing itself through cases, attention to the rhetoric of narrative is necessary in order to remain vigilant to the fact that all cases attempt to impose a perspective upon the reader.

What needs to be discovered is not some innocuous way to write cases but a series of readings of ethics cases that uncovers the rhetorical force of the case. If cases are the data for bioethics, we must come to understand how our data is rhetorically shaped, not so that unbiased cases can be written but so that we can see how the case's narrative elements thwart challenges to the author's philosophy. What I propose is that we do not so much need thicker or richer cases (as so many have advocated) as more sophisticated readings of cases. Reading cases with attention to their fictional qualities, that is, their constructedness, in turn reveals how dilemmas are framed in ways that conceal as well as reveal other ways of seeing. I do not intend this to mean that we are helpless victims of our worldview, condemned to write only that which our moral frames allow, for I hope that the writers of cases become their own critical readers as well. To ignore the narrative characteristics that the bioethics case shares with fiction is to confuse representation with the thing it represents—to mistake the story with the reality—and thus to miss the theory in the case.

References

Dax's Case. A Demand To Die, 1985.

A Right To Die?: The Case of Dax Cowart. Carnegie Mellon University, Center for Design of Educational Computing, 1990.

Abbott, Craig S. "The Case of Debbie Revisited: A Literary Perspective." *Journal of Medical Humanities* 10, no. 2. Fall/Winter (1989): 99–106.

Abrams, Natalie, and Michael D. Buckner, eds. *Medical Ethics: A Clinical Textbook and Reference for the Health Care Professions*. Cambridge, Massachusetts: The MIT Press, 1983.

Ackerman, Terrence F., and Carson Strong. *A Casebook of Medical Ethics*. New York: Oxford University Press, 1989.

Annas, George, and Joseph M. Healey, Jr. "The Patient Rights Advocate: Redefining the Doctor-Patient Relationship in the Hospital Context." *Medical Ethics*. Ed. Natalie Abrams and Michael Buckner. Cambridge: MIT Press, 1983. 211–19.

Anspach, Renee R. "Notes on the Sociology of Medical Discourse: The Language of Case Presentation." *Journal of Health and Social Behavior* 29 (1988): 357–75.

Arras, John D. "Getting Down to Cases: The Revival of Casuistry in Bioethics." *Journal of Medicine and Philosophy* 16 (1991): 29–51.

Bakhtin, Mikhail. "Forms of Time and of the Chronotope in the Novel." *The Dialogic Imagination.* Ed. Michael Holquist. Trans. Caryl Emerson and Michael Holquist. Austin: University of Texas Press, 1981. 84–258.

Bal, Mieke. *Narratology: An Introduction to the Theory of Narrative.* Trans. Christine van Boheemen. London: University of Toronto Press, 1985.

Banfield, Ann. *Unspeakable Sentences: Narration and Representation in the Language of Fiction.* Boston: Routledge, 1992.

Barrett, Edward, ed. *Text, Context, and Hypertext.* Cambridge: MIT Press, 1988.

Barthes, Roland. "From Work to Text." *Image/Music/Text.* Trans. Stephen Heath. New York: Farrar, Straus, and Giroux, 1977. 155–64.

——*Image Music Text*. Trans. Stephen Heath. New York: Farrar, Straus and Giroux, 1992.

——"The Reality Effect." *The Rustle of Language.* Ed. Francois Wahl. Trans. Richard Howard. Berkeley: University of California Press, 1986. 141–48.

——*S/Z*. Trans. Richard Miller. New York: Hill and Wang, 1974.

Bateson, Gregory. *Steps to an Ecology of Mind.* New York: Ballantine Books, 1972.

Bauman, Richard. *Story, Performance, and Event: Contextual Studies of Oral Narrative.* Cambridge: Cambridge University Press, 1986.

Beauchamp, Tom L., and James F. Childress. *Principles of Biomedical Ethics.* 3rd ed. New York and Oxford: Oxford University Press, 1989.

——*Principles of Biomedical Ethics.* 4th ed. New York: Oxford University Press, 1994.

Beauchamp, Tom L., and Laurence B. McCullough. *Medical Ethics: The Moral Responsibilities of Physicians*. Englewood Cliffs, New Jersey: Prentice-Hall, 1984.

Becker, Howard S. *Writing for Social Scientists*. Chicago: University of Chicago Press, 1986.

Berk, Emily, and Joseph Devlin, eds. *Hypertext/ Hypermedia Handbook*. New York: McGraw-Hill Publishing Company, 1991.

Booth, Wayne C. *Critical Understanding: The Powers and Limits of Pluralism*. Chicago: University of Chicago Press, 1979.

——*The Rhetoric of Fiction*. 2nd ed. Chicago: University of Chicago Press, 1983.

Brody, Baruch A. *Life and Death Decision Making*. New York: Oxford University Press, 1988.

Brody, Baruch A., and H. Tristram Engelhardt, Jr. *Bioethics: Readings and Cases*. Englewood Cliffs, N.J.: Prentice-Hall, 1987.

Brody, Howard. *Ethical Decisions in Medicine*. Boston: Little, 1981.

——*The Healer's Power*. New Haven: Yale University Press, 1992.

——*Stories of Sickness*. New Haven: Yale University Press, 1988.

Burrell, David, and Stanley Hauerwas. "From System to Story: An Alternative Pattern for Rationality in Ethics." *Knowledge, Value and Belief*. Ed. Jr. and Daniel Callahan H. Tristram Engelhardt. Hastings-on-Hudson, N.Y.: Hastings Center, 1977. 111–52. Vol. 2.

Burt, Robert A. *Taking Care of Strangers: The Rule of Law in Doctor-Patient Relations*. New York: Free Press, 1979.

Burton, Keith. "A Chronicle: Dax's Case As It Happened." *Dax's Case: Essays in Medical Ethics and Human Meaning*. Ed. Lonnie D. Kliever. Dallas:

Southern Methodist University Press, 1989. 1–12.

Carse, Alisa L. "The 'Voice of Care': Implications for Bioethical Education." *The Journal of Medicine and Philosophy* 16 (1991): 5–28.

Chandler, Raymond. *The Midnight Raymond Chandler*. Boston: Houghton Mifflin, 1971.

Charon, Rita. "The Case: A Relative Stranger." *Second Opinion* 16. March (1991): 50–56.

——"Narrative Contributions to Medical Ethics." *A Matter of Principles: Ferment in U.S. Bioethics*. Ed. R. P. Hamel E. R. Dubose, and L. J. O'Connell. Valley Forge: Trinity Press, 1994. 260–83.

Charon, Rita, et al. "Literature and Medicine: Contributions to Clinical Practice." *Annals of Internal Medicine* 122/8.15 April (1995): 599–606.

Chatman, Seymour. *Coming to Terms: The Rhetoric of Narrative in Fiction and Film*. Ithaca: Cornell University Press, 1990.

——*Story and Discourse*. Ithaca: Cornell University Press, 1978.

——"What Can We Learn from Contextualist Narratology." *Poetics Today* 11 (1990): 309–28.

Childress, James. *Who Should Decide? Paternalism in Health Care*. New York: Oxford University Press, 1982.

——"The Normative Principles of Medical Ethics." *Medical Ethics*. Ed. Robert M. Veatch. Boston: Jones and Bartlett Publishers, 1989. 27–47.

Cixous, Hélène, and Catherine Clement. *The Newly Born Woman*. Trans. Betsy Wing. Minneapolis: University of Minnesota Press, 1986.

Clouser, K. Danner. "Bioethics and Philosophy." *Hastings Center Report*. November–December (1993): S10–S11.

References

——"Philosophy, Literature, and Ethics: Let the Engagement Begin." *Journal of Medicine and Philosophy* 21, No. 3. June (1996): 321–40.

Cohn, Dorrit. *Transparent Minds: Narrative Modes for Presenting Consciousness in Fiction*. Princeton, New Jersey: Princeton University Press, 1978.

Conquergood, Dwight. "Rethinking Ethnography: Towards a Critical Cultural Politics." *Communication Monographs* 58.2 (1991): 179–94.

Cranny-Francis, Anne. *Feminist Fiction: Feminist Uses of Generic Fiction*. Oxford: Polity Press, 1990.

Crigger, Bette-Jane. "Twenty Years After: The Legacy of the Tuskeegee Syphilis Study." *Hastings Center Report*. November–December (1992): 29.

Crysdale, Cynthia S. W. "Gilligan and the Ethics of Care: An Update." *Religious Studies Review* 20/1. January (1984): 21–28.

Culler, Jonathan. *Structuralist Poetics: Structuralism, Linguistics, and the Study of Literature*. Ithaca: Cornell University Press, 1975.

Culver, Charles M., et al. "Basic Curricular Goals in Medical Ethics." *The New England Journal of Medicine* 312/4. Jan. 24 (1985): 253–56.

Davis, Dena S. "Rich Cases: The Ethics of Thick Description." *Hastings Center Report* 21.4 (1991): 12–17.

Delany, Paul, and George P. Landow, eds. *Hypermedia and Literary Studies*. Cambridge: MIT Press, 1990.

Donnelly, William J. "Hypothetical Case Histories: Stories Neither Fact Nor Fiction." Unpublished manuscript.

Donovan, Josephine. "Sarah Orne Jewett's Critical Theory: Notes Toward a Feminine Literary Mode." *Critical Essays on Sarah Orne Jewett*. Ed. Gwen L. Nagel. Boston: G. K. Hall, 1984. 212–25.

Douglas, J. Yellowlees. "'How Do I Stop This Thing?': Closure and Interdeterminacy in Interactive Narratives." *Hyper/Text/Theory*. Ed. George P. Landow. Baltimore: The Johns Hopkins University Press, 1994. 159–88.

Elliott, Carl. "Philosopher Assisted Suicide and Euthanasia." *British Medical Journal* 313.26 October (1996): 1088–889.

Engelhardt, H. Tristram. *The Foundations of Bioethics*. New York: Oxford University Press, 1986.

Fletcher, John C., Norman Quist, and Albert R. Jonsen, eds. *Ethics Consultation in Health Care*. Ann Arbor, Michigan: Health Administration Press, 1989.

Forster, E. M. *Aspects of the Novel*. New York: Harcourt, 1927.

Frank, Arthur. *The Wounded Storyteller*. Chicago: University of Chicago Press, 1995.

Geertz, Clifford. *Works and Lives*. Stanford: Stanford University Press, 1988.

Genette, Gérard. *Narrative Discourse*. Trans. Jane E. Lewin. Ithaca, New York: Cornell University Press, 1980.

———*Paratexts: Thresholds of Interpretation*. Trans. Jane E. Lewin. Cambridge: Cambridge University Press, 1997.

Gert, Bernard, and Charles M. Culver. *Philosophy in Medicine: Conceptual and Ethical Issues in Medicine and Psychiatry*. New York: Oxford University Press, 1982.

Goffman, Erving. *Frame Analysis*. Cambridge: Harvard University Press, 1974.

Goody, Jack, and Ian Watt. "The Consequences of Literacy." *Literacy in Traditional Societies*. Ed. Jack Goody. Cambridge: Cambridge University Press, 1968. 27–68.

Greimas, A. J., and Joseph Courtés. *Semiotics and Language: An Analytical Dictionary.* Trans. Larry Crist. Bloomington: Indiana University Press, 1982.

Hauerwas, Stanley. "Towards an Ethics of Character." *Vision and Virtue: Essays in Christian Ethical Reflection.* Nortre Dame, Indiana: Fides Publisher, 1971.

——*Vision and Virtue: Essays in Christian Ethical Reflection.* Nortre Dame, Indiana: Fides Publishers, 1974.

Havelock, Eric A. *Preface To Plato.* Cambridge: Harvard University Press, 1963.

Hawkins, Anne Hunsaker. "Literature, Medical Ethics, and 'Epiphanic Knowledge.'" *Journal of Clinical Ethics* 5/4. Winter (1994): 283–90.

Hawkins, Anne Hunsaker. "Literature, Philosophy, and Medical Ethics: Let the Dialogue Go On." *Journal of Medicine and Philosophy* 21, No. 3. June (1996): 341–54.

Highsmith, Patricia. *The Talented Mr. Ripley.* In *Crime Novels: American Noir of the 1950s.* New York: Library of America, 1997. 161–404.

Hunter, Kathryn Montgomery. *Doctors' Stories: The Narrative Structure of Medical Knowledge.* Princeton, New Jersey: Princeton University Press, 1991.

——"Narrative, Literature, and the Clinical Exercise of Practical Reason." *Journal of Medicine and Philosophy* 21 (1996): 303–20.

——"Overview." *Second Opinion* 15. November (1990): 64–67.

Iser, Wolfgang. *The Implied Reader: Patterns of Communication in Prose Fiction from Bunyan to Beckett.* London: The Johns Hopkins University Press, 1974.

James, Henry. "The Art of Fiction." *Essential of the*

Theory of Fiction. Ed. Michael J. Hoffman and Patrick D. Murphy. 2nd ed. Durham: Duke University Press, 1996. 14–21.

Jones, Anne Hudson. "Darren's Case: Narrative Ethics in Perri Klass's *Other Women's Children.*" *Journal of Medicine and Philosophy* 21/3. June (1996): 267–86.

Jones, Anne Hudson. "Literature and Medicine: Narrative Ethics." *Lancet* 349.26 April (1997): 1243–46.

Jonsen, Albert R. "Casuistry as Methodology in Clinical Ethics." *Theoretical Medicine* 12 (1991): 395–407.

Jonsen, Albert R., and Andrew Jameton. "History of Medical Ethics: The United States in the Twentieth Century." *Encyclopedia of Bioethics*. Ed. Warren T. Reich. Revised ed. New York: Simon & Schuster, 1995. 1616–32. Vol. 3.

Jonsen, Albert R., Mark Siegler, and William J. Winslade. *Clinical Ethics*. 4th ed. New York: McGraw-Hill, 1998.

Joyce, Michael. *Afternoon, a story*. Computer software. Eastgate Systems, 1987.

Kermode, Frank. *The Sense of an Ending: Studies in the Theory of Fiction*. New York: Oxford University Press, 1967.

Kilner, John F. *Who Lives? Who Dies?* New Haven: Yale University Press, 1990.

Kleinman, Arthur. *Writing at the Margin: Discourse Between Anthropology and Medicine*. Berkeley: University of California Press, 1995.

Kliever, Lonnie. "Preface." *Dax's Case: Essays in Medical Ethics and Human Meaning*. Ed. Lonnie Kliever. Dallas: Southern Methodist University Press, 1989. xi–xvii.

Kopelman, Loretta M. "Moral Problems in Psychiatry." *Medical Ethics*. Ed. Robert M. Veatch. Boston: Jones and Bartlett, 1989. 253–90.

References

Kristeva, Julia. *Desire in Language: A Semiotic Approach to Literature and Art.* Ed. Leon Roudiez. Oxford, 1980.

Landow, George P., ed. *Hyper/Text/Theory.* Baltimore: The Johns Hopkins University Press, 1994.

———*Hypertext.* Ed. Gerald Prince, Stephen G. Nichols, and Wendy Steiner. Baltimore: The Johns Hopkins University Press, 1992.

Langer, Susanne. "A Note on the Film." *Film: A Montage of Theories.* Ed. Richard Dyer MacCann. New York: E. P. Dutton, 1966. 199–204.

Lanser, Susan Sniader. *The Narrative Act: Point of View in Prose Fiction.* Princeton: Princeton University Press, 1981.

———"Toward a Feminist Narratology." *Essential of the Theory of Fiction.* Ed. Michael J. Hoffman and Patrick D. Murphy. 2nd ed. Durham: Duke University Press, 1996. 453–72.

Levine, Carol, and Robert Veatch, eds. *Cases in Bioethics from the Hastings Center Report.* Hastings-on-Hudson, NY: The Hastings Center, 1982.

Levine, Melvin D., Lee Scott, and William J. Curran. "Ethics Rounds in a Children's Medical Center: Evaluation of a Hospital-Based Program for Continuing Education in Medical Ethics." *Pediatrics* 60. August (1977): 205.

Liestol, Gunnar. "Wittgenstein, Genette, and the Reader's Narrative in Hypertext." *Hyper/Text/Theory.* Ed. George P. Landow. Baltimore: Johns Hopkins University Press, 1994. 87–120.

Lodge, David. *The Art of Fiction.* New York: Penguin, 1992.

Lotman, Jurij. *The Structure of the Artistic Text.* Trans. Ronald Vronn. Ann Arbor: University of Michigan Press, 1977.

Macklin, Ruth. *Mortal Choices: Bioethics in Today's World.* New York: Pantheon, 1987.

Martin, Wallace. *Recent Theories of Narrative*. Ithaca: Cornell University Press, 1986.

May, William F. *The Physician's Covenant: Images of the Healer in Medical Ethics*. Philadelphia: The Westminster Press, 1983.

McCloud, Scott. *Understanding Comics*. New York: Harper, 1993.

McCullough, Laurence B., and Frank A. Chervenak. *Ethics in Obstetrics and Gynecology*. New York: Oxford University Press, 1994.

McHale, Brian. "Free Indirect Discourse: A Survey of Recent Accounts." *Poetics and Theory of Literature* 3 (1978): 249–87.

McLuhan, Marshal. *Understanding Media: The Extensions of Man*. Cambridge: The MIT Press, 1964.

Miles, Steven, and Kathryn Montgomery Hunter. "Case Stories." *Second Opinion* 15 (1990): 60–69.

Miller, Franklin G., Joseph J. Fins, and Matthew D. Bacchetta. "Clinical Pragmatism: John Dewey and Clinical Ethics." *Journal of Contemporary Health Law and Policy* 13 (1996): 27–51.

Mintz, David. "What's in a Word: The Distancing Function of Language in Medicine." *The Journal of Medical Humanities* 13.4 (1992): 223–33.

Mitchell, W. J. T. "Representation." *Critical Terms for Literary Study*. Ed. Frank Lentricchia and Thomas McLaughlin. Chicago: University of Chicago Press, 1995. 11–22.

Morreim, E. Haavi. *Balancing Act: The New Medical Ethics of Medicine's New Economics*. Washington, D.C.: Georgetown University Press, 1995.

Morson, Gary Saul. "The Reader as Voyeur." *Tolstoy's Short Fiction*. Ed. Michael R. Katz. New York: Norton, 1991. 379–94.

Morson, Gary Saul, and Caryl Emerson. *Mikhail Bakhtin: Creation of a Prosaics*. Stanford: Stanford University Press, 1990.

Nelson, Hilde Lindemann, ed. *Stories and Their Limits*. New York: Routledge, 1997.

Nelson, James Lindemann. "Measured Fairness, Situated Justice: Feminist Reflections on Health Care Rationing." *Kennedy Institute of Ethics Journal* 6, No. 1 (1996): 53–63.

Ong, Walter J. *Orality and Literacy: The Technologizing of the Word*. Ed. Terence Hawkes. London and New York: Methuen, 1986.

Pappas, Gregory F. "Dewey's Ethics: Morality as Experience." *Reading Dewey*. Ed. Larry A. Hickman. Bloomington: Indiana University Press, 1998. 100–23.

Pellegrino, Edmund. "To Look Feelingly—The Affinities of Medicine and Literature." *Literature and Medicine* 1 (1982): 18–22.

Pellegrino, Edmund D. "The Metamorphosis of Medical Ethics: A 30-Year Retrospective." *Journal of the American Medical Association* 269.9 (1993): 1158–62.

Pellegrino, Edmund D., and David C. Thomasma. *The Virtues in Medical Practice*. New York: Oxford University Press, 1993.

Perrin, Norman. *What Is Redaction Criticism?* Ed. Dan O. Via. Philadelphia: Fortress Press, 1978.

Prince, Gerald. *Dictionary of Narratology*. Lincoln: University of Nebraska Press, 1987.

———"Introduction to the Study of the Narratee." *Reader-Response Criticism: From Formalism to Post-Structuralism*. Ed. Jane P. Tompkins. Baltimore: Johns Hopkins University Press, 1980. 7–25.

———"Narratology, Narratological Criticism, and Gender." *Fiction Updated: Theories of Fictionality, Narratology, and Poetics*. Ed. Calin-Andrei Mihailescu and Walid Hamarneh. Toronto: University of Toronto Press, 1996. 159–64.

Quill, Timothy. "Death and Dignity: A Case of

Individualized Decision Making." *New England Journal of Medicine* 324 (1991): 691–94.

Radwany, S. M., and B. H. Adelson. "The Use of Literary Classics in Teaching Medical Ethics." *JAMA* 257 (1987): 1629–30.

Reich, Warren Thomas. "Caring for Life in the First of It: Moral Paradigms for Perinatal and Neonatal Ethics." *Seminars in Perinatology* 11/3. July (1987): 279–87.

Ricoeur, Paul. *Time and Narrative.* Vol. 2. Chicago: University of Chicago Press, 1984.

Rimmon-Kenan, Shlomith. *Narrative Fiction: Contemporary Poetics.* London: Routledge, 1983.

Rothman, David J. *Strangers at the Bedside*: Basic Books, 1991.

Schoene-Seifert, Bettina, and James F. Childress. "How Much Should the Cancer Patient Know and Decide?" *Ca-A Cancer Journal for Clinicians* 36/2. March/April (1986): 85–95.

Scholes, Robert. *Structuralism in Literature: An Introduction.* New Haven and London: Yale University Press, 1974.

Sherwin, Susan. "A Feminist Approach to Ethics." *Dalhousie Review* 64 (1987): 704–13.

Sherwin, Susan. *No Longer Patient: Feminist Ethics and Health Care.* Philadelphia: Temple University Press, 1992.

Slatin, John. "Reading Hypertext." *Hypermedia and Literary Studies.* Ed. Paul Delany and George P. Landow. Cambridge, MA: The MIT Press, 1991.

Smith, Barbara Herrnstein. *Poetic Closure: A Study of How Poems End.* Chicago: University of Chicago Press, 1968.

Stam, Robert, Robert Burgoyne, and Sandy Flitterman-Lewis. *New Vocabularies in Film Semiotics.* London: Routledge, 1992.

Terry, James S., and Peter C. Williams. "Literature and

Bioethics: The Tension in Goals and Styles."
Literature and Medicine 7 (1988): 1–21.

Thomasma, David C., and Patricia A. Marshall.
Clinical Medial Ethics Cases and Readings.
Lanham: University Press of America, 1995.

Thomson, Judith Jarvis. "A Defense of Abortion."
Classic Works in Medical Ethics. Ed. Gregory E.
Pence. Boston: McGraw-Hill, 1998. 153–68.

Todorov, Tzvetan. *The Fantastic.* Ithaca: Cornell
University Press, 1975.

———*The Poetics of Prose.* Trans. Richard Howard.
Ithaca: Cornell University Press, 1977.

Toolan, Michael J. *Narrative: A Critical Linguistic
Introduction.* London: Routledge, 1988.

Toulmin, Stephen. "How Medicine Saved the Life of
Ethics." *Perspectives in Biology and Medicine*
25 (1982): 736–50.

Turner, Victor, and Edie Turner. "Performing Ethnography."
The Anthropology of Performance. New York:
PAJ Publications, 1987. 139–55.

Uspensky, Boris. *A Poetics of Composition.* Trans.
Valentina Zavarin and Susan Wittig. Berkeley:
University of California Press, 1973.

Veatch, Robert. *A Theory of Medical Ethics.* New York:
Basic Books, 1981.

Warshaw, Carole, and Suzanne Poirier. "Hidden Stories
of Women." *Second Opinion* 17, no. 2. October
(1991): 48–61.

Waymack, Mark H., and George A. Taler. *Medical
Ethics and the Elderly: A Casebook.* Chicago:
Pluribus Press, 1988.

White, Robert B. "A Demand to Die." *Hastings Center
Report* 5. June (1975): 9–10.

———"A Memoir: Dax's Case Twelve Years Later." *Dax's
Case: Essays in Medical Ethics and Human
Meaning.* Ed. Lonnie D. Kliever. Dallas: Southern
Methodist University Press, 1989. 13–22.

Zaner, Richard M. *Ethics and the Clinical Encounter*.
Englewood Cliffs, New Jersey: Prentice, 1988.

Permissions

Index